SAE International's Dictionary of Electric and Hybrid Vehicles

SAE International's Dictionary of Electric and Hybrid Vehicles

JON M. QUIGLEY
FRED STARKEY

INTERNATIONAL®

Warrendale, Pennsylvania, USA

400 Commonwealth Drive
Warrendale, PA 15096-0001 USA
E-mail: CustomerService@sae.org
Phone: 877-606-7323 (inside USA and Canada)
724-776-4970 (outside USA)
FAX: 724-776-0790

Library of Congress Catalog Number 2024944284
http://dx.doi.org/10.4271/9781468608533

ISBN-Print 978-1-4686-0852-6
ISBN-PDF 978-1-4686-0853-3
ISBN-epub 978-1-4686-0854-0

To purchase bulk quantities, please contact: SAE Customer Service

E-mail: CustomerService@sae.org
Phone: 877-606-7323 (inside USA and Canada)
724-776-4970 (outside USA)
Fax: 724-776-0790

Visit the SAE International Bookstore at books.sae.org

Publisher
Sherry Dickinson Nigam

Product Manager
Amanda Zeidan

Production and Manufacturing Associate
Michelle Silberman

Contents

Preface

Welcome to *SAE International's Dictionary of Electric and Hybrid Vehicles*. As electric and hybrid vehicles continue to transform the automotive industry, the technical language and concepts surrounding them grow increasingly complex. This dictionary aims to demystify the terminology associated with these cutting-edge vehicles, offering clear and concise definitions to enhance understanding for engineers, students, professionals, and enthusiasts alike. Whether you are navigating the intricacies of battery management systems or exploring the latest advancements in regenerative braking, this resource will serve as a comprehensive guide to the evolving world of electric and hybrid vehicle technology.

The dictionary encompasses many terms, from basic principles to more complex concepts covering various aspects of electric and hybrid vehicle technologies. Experts carefully curated and reviewed each entry to ensure accuracy and relevance. The definitions are concise yet comprehensive, providing clear explanations and context that enhance understanding and application.

This dictionary is the result of extensive collaboration among leading experts in automotive engineering. It reflects the latest advancements and trends in the field. It includes terminology from electrical engineering, mechanical engineering, chemistry, and environmental science, making it a holistic resource.

Our goal is to provide a reliable and up-to-date resource that supports the ongoing learning and development of professionals and students. Whether you are a seasoned engineer looking to refresh your knowledge base or a student embarking on your first project, this dictionary aims to be your go-to source for clear and authoritative information. However, the rate of change in technology and application in the electric vehicle (EV) market makes this a challenging endeavor.

We hope this dictionary will assist you in mastering the technical language and inspire further innovation and exploration in electric and hybrid vehicles. Your journey through the dynamic world of automotive technology starts here.

As we conclude this preface, it is important to note that you will encounter many acronyms throughout this book, some of which may even be the same three or four letters but have different meanings depending on the context. This illustrates a common challenge in communication; in verbal exchanges, we can clarify by asking what a specific acronym means. For a detailed explanation of all the acronyms used, please see the List of Acronyms.

Acknowledgments

We want to start by thanking Sherry Nigam for starting these dictionary projects and Amanda Zeidan for the many hours of reviews and feedback that drove the result from SAE International for the project, and all their efforts in this and the previous dictionary works. This book would not be here if not for them. Nevertheless, any defects or anomalies are solely the responsibility of the authors.

Next, we thank those who have read our business, project management, and product development books. Sometimes, we wonder if these books have the positive impact we wish for them. This is why we write a book. One does not write business and technical books to become rich; let us dispel that now if you have thoughts to the contrary. We recently had a couple of phone calls where readers have told us how they have used the books and made things better, providing tangible descriptions of that improvement.

Lastly, many people have helped us get where we are; I mention some key people, starting with Tony Singleton, an old-school engineer who shared what he knew with a fresh out-of-school engineering student. Next, I want to recognize E. Allen Downs, another person key in my career development; he introduced me to the automotive product development industry on a much larger scale. I am also eternally grateful for how he handled a restructuring event at the company where I worked, which resulted in my dismissal. However, I was hired back within a few weeks when the executives realized they made a mistake, this was due to the efforts of E. Allen Downs and Fred Starkey. The list of who I owe is long, but a couple of more names include Rick Bertalan for bringing me into commercial vehicles, and Jerry Obermeyer for showing me that failure is not terminal and that failing usually means we are trying to do something significant and of value.

Introduction

Welcome to *SAE International's Dictionary of Electric and Hybrid Vehicles.* This comprehensive resource is designed to serve as a definitive guide for students, engineers, policymakers, and enthusiasts navigating the rapidly evolving landscape of electric and hybrid vehicle technologies. As the automotive industry undergoes a significant transformation fueled by technological advances and shifting global demands for sustainability, the vocabulary and concepts associated with these changes continue to expand and evolve.

This dictionary captures the essence of these dynamic developments, offering detailed definitions and explanations of terms, technologies, and concepts critical to understanding and advancing the field of electric and hybrid vehicles. From foundational elements like battery chemistries and motor types to cutting-edge innovations in autonomous driving and smart charging networks, this dictionary is structured to define terms and contextualize them within the larger framework of modern automotive engineering and environmental impact.

As technology progresses and new applications are discovered, this dictionary will serve as a vital tool for keeping pace with the new terminology and innovations shaping transportation's future. Whether you are a veteran automotive engineer or a curious newcomer, *SAE International's Dictionary of Electric and Hybrid Vehicles* aims to enrich your understanding and appreciation of this exciting and important technology area.

List of Acronyms

AACHS	Active Air Cooling and Heating Systems
AC	Alternating Current
ACD	Automatic Connection Device
ACS	Accelerator-Controlled System
AER	All-Electric Range
AFC	Alkaline Fuel Cell
Ah	Ampere-Hour
AID	Actuation of Isolation Detection
APU	Auxiliary Power Unit
AVAS	Acoustic Vehicle Alerting System
AWD	All-Wheel Drive
Back EMF	Back Electromotive Force
BEV	Battery Electric Vehicle
BLDC	Brushless DC Motor
BMS	Battery Management Systems
BSS	Battery Swapping Station
CaaS	Charging as a Service
CAN	Controller Area Network
CARB	California Air Resources Board
CATV	Community Antenna Television
CBET	Charge Before End Time
CCI	Charge Circuit Interrupter
CCID	Charge Circuit Interrupting Device
CCS	Combined Charging System
CD	Charge-Depleting
CEMS	Customer Energy Management Service
CO_2	Carbon Dioxide
CPO	Charge Point Operator
CPP	Critical Peak Pricing
CPUC	California Public Utilities Commission
CPV	Concentrated Photo Voltaic
CSA	Connectivity Standards Alliance

CVT	Continuously Variable Transmission
DC	Direct Current
DER	Distributed Energy Resource
DG	Distributed Generation
DHCP	Dynamic Host Configuration Protocol
DME	Distributed Managing Entity
DMS	Distribution Management System
DoD	Depth of Discharge
DOT	Department of Transportation
DR	Demand Response
DRFPI	Direct Radio Frequency Power Injection
DTCs	Diagnostic Trouble Codes
DWPT	Dynamic Wireless Power Transfer
DyFeB	Dysprosium–Iron–Boron
ECU	Electronic Control Unit
EMC	Electromagnetic Compatibility
EMF	Electromagnetic Field
EMI	Electromagnetic Interference
EMS	Energy Management System
EPCU	Electric Power Control Unit
EPL	Electrical Physical Layer
EPR	Ethylene Propylene Rubber
EPS	Electric Power Steering
EREV	Extended-Range Electric Vehicle
ESC	Energy Service Certificate
ESD	Electrostatic Discharge
ESS	Energy Storage System
EU	European Union
EUMD	Electric Vehicle Usage and Maintenance Data
EV	Electric Vehicle
EV MVLL	Electric Vehicle Maximum Voltage Low Limit
EVCC	Electric Vehicle Communication Controller
EVHS	Electric Vehicle Home-charge Scheme
EVSE	Electric Vehicle Service Equipment
EVSE	Electric Vehicle Supply Equipment
EVSP	Electric Vehicle Service Provider
EVT	Electric Variable Transmission
FCHEV	Fuel Cell Hybrid Electric Vehicle
FCV	Fuel Cell Vehicle
G2V	Grid to Vehicle
GA	Ground Assembly
GaN	Gallium Nitride

GFCIs	Ground-Fault Circuit Interrupters
GHG	Greenhouse Gas
GM	General Motors
GPS	Global Positioning System
HDEV	Hybrid Diesel Electric Vehicle
HEV	Hybrid Electric Vehicle
HPU	Hybrid Power Unit
HV	Hybrid Vehicles
HVAC	Heating, Ventilation, and Air Conditioning
HVBD	Hazardous Voltage Bus Discharge
HVD	High Voltage Disconnect
HVIL	High-Voltage Interlock Loop
HVIS	High-Voltage Interlock Systems
ICNIRP	International Commission on Non-Ionizing Radiation Protection
ICs	Integrated Circuits
ICE	Internal Combustion Engine
IDC	Indian Driving Cycle
IEC	International Electrotechnical Commission
IEEE	Institute of Electrical and Electronics Engineers
IGBTs	Insulated Gate Bipolar Transistors
IMD	Insulation Monitoring Devices
IMO	International Maritime Organization
IoT	Internet of Things
IP	Internet Protocol
IP64	Ingress Protection-64
IQPC	International Quality & Productivity Centre
ISC	Infrastructure-Side Connection
JEVS	Japan Electric Vehicle Association Standards
kHz	Kilohertz
kW	Kilowatts
kWh	Kilowatt-Hours
LCO	Lithium Cobalt Oxide
LD-PEV	Light-Duty Plug-In Electric Vehicle
LEDs	Light-Emitting Diodes
LFP	Lithium Iron Phosphate
$LiCoO_2$	Lithium Cobalt Oxide
$LiFePO_4$	Lithium Iron Phosphate
Li-ion	Lithium-Ion
$LiMn_2O_4$	Lithium Manganese Oxide
$LiPF_6$	Lithium Hexafluorophosphate

LMO	Lithium Manganese Oxide
LR-WPANs	Low-Rate Wireless Personal Area Network
MCC	Motor Control Center
MHz	Megahertz
MIDC	Modified Indian Driving Cycle
Mi/h	Miles per hour
mi/kWh	Miles per Kilowatt-Hours
Mild HV	Mild Hybrid Vehicle
MJ	Megajoules
MOSFETs	Metal-Oxide-Semiconductor Field-Effect Transistors
MPG	Miles per Gallon
MPGe	Miles per Gallon Equivalent
MQTT	Message Queuing Telemetry Transport
MVLL	Maximum Voltage Low Limit
MWh	Megawatt-Hour
NACS	North American Charging System
NdFeB	Neodymium Iron Boron
NEDC	New European Driving Cycle
NEV	Neighborhood Electric Vehicle
NFC	Near-Field Communication
NHEV-TSA	National Highway Electric Vehicle Traffic Safety Administration
NHTSA	National Highway Traffic Safety Administration
NiCd	Nickel-Cadmium
NiMH	Nickel-Metal Hydride
NiOOH	Nickel Oxyhydroxide
NiZn	Nickel-Zinc
Nm	Newton-Meters
NO_2	Nitrogen Dioxide
OBC	On-Board Charger
OBD	Onboard Diagnostics
OBRESS	On-Board Rechargeable Energy Storage System
OCPP	Open Charge Point Protocol
OEM	Original Equipment Manufacturer
OpenADR	Open Automated Demand Response
OTA	Over the Air
OVCC	Off-Vehicle Charge Capable
OZEV	Office for Zero Emission Vehicles
PEC	Power Electronics Controller
PEMFC	Proton Exchange Membrane Fuel Cell
PEV	Plug-In Electric Vehicle
PHEV	Plug-In Hybrid Electric Vehicle

PKI	Public Key Infrastructure
PLC	Power Line Communication
PM	Particulate Matter
PMSM	Permanent Magnet Synchronous Motor
PPE	Personal Protective Equipment
PrFeB	Praseodymium Iron Boron
PSD	Auto-Power Spectral Density
PST	Power-Split Transmission
PV	Photovoltaic
PVC	Polyvinyl Chloride
PWM	Pulse Width Modulation
PWM-CP	Pulse Width Modulation—Control Pilot
PZEV	Partial Zero Emission Vehicle
RECs	Renewable Energy Credits
RESS	Rechargeable Energy Storage System
RESS	Remote Energy Storage System
RESS Launch	Rechargeable Energy Storage System Launch
RF	Radiofrequency
RFID	Radiofrequency Identification
RPM	Revolutions per Minute
RRC	Rolling Resistance Coefficient
RWD	Rear-Wheel Drive
SCs	Supercapacitors
SEP	Smart Energy Profile
SEP 2.0	Smart-Energy Profile 2.0
SESM	Separately Excited Synchronous Motor
SGIP	Smart Grid Interoperability Panel
SHEV	Series Hybrid Electric Vehicle
SI	International System of Units
SiC	Silicon Carbide
SmCo	Samarium-Cobalt
SOA	Safe Operating Area
SOC	State of Charge
SOFC	Solid Oxide Fuel Cell
SOH	State of Health
SPaMCAST	Software Process and Measurement Cast
SRM	Switched Reluctance Motors
SSR	Solid-State Relay
SULEV	Super-Ultra-Low Emission Vehicle
TCP	Transmission Control Protocol
TEMCH	Transmission Energy Market Clearing House
TEPCO	Tokyo Electric Power Company

TLS	Transport Layer Security
TOU	Time of Use
UBE	Usable Battery Energy
UDDS	Urban Dynamometer Driving Schedule
UDP	User Datagram Protocol
UL	Underwriters Laboratories
ULEZ	Ultra-Low-Emission Zone
UNECE	United Nations Economic Commission for Europe
UPS	Uninterruptible Power Supplies
US	United States
V2B	Vehicle-to-Building
V2G	Vehicle-to-Grid
V2G AC	Vehicle-to-Grid via Alternating Current
V2G DC	Vehicle-to-Grid via Direct Current
V2H	Vehicle-to-Home
V2I	Vehicle-to-Infrastructure
V2L	Vehicle-to-Load
V2N	Vehicle-to-Network
V2P	Vehicle-to-Pedestrian
V2V	Vehicle-to-Vehicle
V2X	Vehicle-to-Everything
VA	Vehicle Assembly
VCE	Vehicle Control Electronics
VCU	Vehicle Control Unit
VDC	Volts Direct Current
VFDs	Variable-Frequency Drives
VRFB	Vanadium Redox Flow Batteries
VSC	Vehicle-Side Connection
WCS	Wireless Charging System
Wh	Watt-Hour
Wh/kg	Watt-Hours per Kilogram
Wh/L	Watt-Hours per Liter
WLTP	Worldwide Harmonized Light Vehicles Test Procedure
WP.29	World Forum for Harmonization of Vehicle Regulations
WPT	Wireless Power Transfer
WTW	Well-to-Wheel
xEV	Electrified Propulsion Vehicle
XLPE	Cross-Linked Polyethylene
ZEV	Zero Emission Vehicle

A

AC (Alternating Current) Level 1 Charging

AC Level 1 charging is the lowest power level of electric vehicle (EV) charging, typically using a standard household electrical outlet. This charging method delivers AC power to the EV's onboard charger (OBC) at a low rate, usually around 120 V in North America and 230 V in Europe, with a maximum charging power of up to 2–3 kW. AC Level 1 charging is the slowest option for EVs, providing relatively low charging speeds. It is also used for overnight charging at home or when faster charging options are not readily available. While AC Level 1 charging is convenient and accessible, it may not be suitable for users with higher daily driving needs or for quickly replenishing the battery's charge.

For additional information, see **J3068** and **J1772** [1, 2].

ACD (Automatic Connection Device)

A system that establishes, without direct human intervention, a conductive electrical interface between on-vehicle and off-vehicle components for the purpose of power transfer.

For additional information, see **J1715** [3].

AC Power Charging Levels

AC power charging levels refer to the standardized categories of charging infrastructure used for EVs, primarily determined by the power output and charging speed they offer. AC charging delivers electric power to an EV's battery using AC, which most residential and commercial electrical outlets supply. The AC power is then converted to direct current (DC) by the vehicle's OBC before charging the battery. The charging levels vary based on the power capacity, measured in kilowatts (kW), and directly influence the speed at which an EV's battery can be charged.

AC Power Circuits

AC power circuits in an EV refer to the electrical systems and components designed to manage AC electricity within the vehicle. These circuits play a critical role in energizing the batteries that provide power for various onboard systems, including the electric motor, OBCs, heating, ventilation, and air conditioning (HVAC) systems, and auxiliary devices.

AC Power Line Electric Fields

AC power line electric fields are those generated by AC power transmission and distribution lines. These electric fields emanate from the electrical conductors carrying AC electricity and extend into the surrounding environment. In the context of EVs, AC power line electric fields may be encountered during charging at locations where EV charging infrastructure is connected to the grid.

Exposure to AC power line electric fields is a concern primarily due to potential health effects associated with electromagnetic fields (EMFs). While scientific research on the health impacts of AC power line electric fields is ongoing, some studies suggest a possible association with increased risk of certain health conditions, although conclusive evidence is lacking.

Regulatory bodies, such as the International Commission on Non-Ionizing Radiation Protection (ICNIRP) and the Institute of Electrical and Electronics Engineers (IEEE), have established guidelines and safety limits for exposure to EMFs, including AC power line electric fields, to protect public health and safety [4].

AC Present Indicator

An AC present indicator, commonly found in EVs and EV charging stations, is a visual or audible signal indicating an AC power supply. This indicator alerts users when AC power is available and ready to charge the vehicle's battery. In EVs, the AC present indicator may illuminate the vehicle's dashboard or charging interface when connected to an AC charging station and power flows to the vehicle's OBC. Similarly, in AC charging stations, the AC present indicator may be a light or display that illuminates when the station is powered and ready to provide AC electricity to charge EVs. The AC present indicator plays a crucial role in facilitating the charging process by giving users explicit feedback on the availability of AC power, helping to ensure safe and efficient charging operations.

AC Receptacle

An AC receptacle, an electrical outlet or power socket, is a standardized electrical fixture that supplies AC power to EVs for charging purposes. AC receptacles come in various configurations and voltage ratings, with common types including household outlets (e.g., NEMA 5-15 and NEMA 5-20) and higher-power outlets (e.g., NEMA 14-50). EV owners can connect their vehicles to AC receptacles using charging cables equipped with compatible plugs, allowing for convenient and accessible charging at residential homes, workplaces, public facilities, and other locations equipped with AC power outlets. AC receptacles provide a slower charging rate compared to dedicated EV charging stations but offer flexibility and accessibility for EV charging in a wide range of environments.

Accepted Charge Rate

The accepted charge rate refers to the maximum rate at which an EV's battery can safely accept electrical energy during charging. It is typically measured in kW and represents the optimal charging speed the battery can handle without causing excessive heat buildup or compromising its long-term durability. The accepted charge rate is influenced by various factors, including the battery's chemistry, temperature, state of charge (SOC), and internal management systems. Charging an EV at a rate that exceeds its accepted charge rate may result in reduced charging efficiency, increased battery degradation, and potential safety hazards. Therefore, EV owners must adhere to the recommended charging guidelines provided by the vehicle manufacturer and charging infrastructure providers to ensure safe and efficient charging operations.

Access Cover Interlocks

Access cover interlocks are safety mechanisms to prevent unauthorized access to critical components or compartments within an EV. These interlocks typically consist of sensors, switches, or locking mechanisms integrated into the vehicle's access covers, panels, or doors. When engaged, access cover interlocks prevent the opening or removal of covers or panels that provide access to high-voltage components, such as the battery pack, power electronics, or electrical connectors.

The primary purpose of access cover interlocks is to enhance safety by restricting access to hazardous areas or components within the EV, particularly those associated with high-voltage electrical systems. By preventing unauthorized personnel or individuals without proper training from accessing these areas, access cover interlocks help minimize the risk of electrical shock, arc flash, or other safety hazards.

Access cover interlocks may be activated automatically when the vehicle is powered on or off, or they may require manual activation by authorized personnel using key switches, codes, or electronic authentication systems. In some cases, access cover interlocks may also trigger visual or audible warnings to alert nearby individuals of restricted access.

Accessibility of High-Voltage Terminals

Accessibility of high-voltage terminals refers to the ease and safety with which service technicians or maintenance personnel can access and work on the high-voltage components within an EV for maintenance, repair, or diagnostic purposes. High-voltage terminals typically include connections, connectors, or components associated with the EV's high-voltage electrical system, such as the battery pack, power electronics, motor controllers, and charging systems.

Ensuring high-voltage terminals are in place is essential to facilitate efficient servicing and maintenance of EVs while minimizing the risk of electrical hazards or accidents. Factors influencing accessibility may include the design and layout of the vehicle's high-voltage components, the availability of access panels or service points, and the implementation of safety features such as interlocks, insulation, and warning labels.

Accessibility considerations are essential for service technicians and maintenance personnel performing tasks such as battery inspection, component replacement, or diagnostic testing on EVs. By prioritizing accessibility and incorporating design features that facilitate safe and efficient access to high-voltage terminals, EV manufacturers can enhance serviceability, reduce downtime, and ensure compliance with safety regulations and standards.

ACS (Accelerator-Controlled Systems)

An ACS in an EV is a control mechanism that regulates the vehicle's speed and acceleration in response to input from the accelerator pedal. Unlike traditional internal combustion engine (ICE) vehicles that rely on throttle cables to control engine speed, EVs utilize electronic systems to manage power output and acceleration.

In an EV equipped with an ACS, depressing the accelerator pedal sends signals to the vehicle's electronic control unit (ECU) or powertrain controller. The ECU then adjusts the power output from the electric motor(s) based on the pedal position and driver input, determining the vehicle's speed and acceleration.

Actual Charge Time

Actual charge time refers to the real-world duration required to recharge the battery of an EV from the specified SOC to another. It considers factors such as the charging rate of the charging station, the current state of the vehicle's battery, and any limitations or constraints that may affect the charging process. Actual charge time may vary depending on various factors, including the battery's capacity, charging infrastructure (such as Level 1, Level 2, or DC fast charging), ambient temperature, and the battery's state of health (SOH). Understanding the actual charge time is essential for EV owners to plan their charging sessions effectively and optimize their driving experience.

Actuation of High-Voltage Disconnect

High-voltage disconnect in an EV refers to physically disconnecting the high-voltage electrical system from the vehicle's powertrain components and other electrical systems. This disconnection is typically achieved by operating a mechanical or electromechanical switch or circuit breaker within the EV's electrical architecture.

The high-voltage disconnect is a safety feature that isolates the battery pack from the rest of the vehicle's electrical systems during maintenance, servicing, or an accident or emergency. Disconnecting the high-voltage system minimizes the risk of electric shock and other electrical hazards, ensuring the safety of technicians, first responders, and occupants.

The actuation of the high-voltage disconnect may be manual, requiring physical intervention by a trained operator, or automatic, triggered by sensors or control systems in response to specific conditions or commands. In either case, the primary objective is to quickly and reliably isolate the high-voltage system to prevent electrical hazards and ensure the safety of personnel and bystanders.

Actuation of Short-Circuit Protection

The actuation of short-circuit protection in an EV refers to the activation of safety mechanisms designed to detect and mitigate electrical short circuits within the vehicle's electrical system [5]. A short circuit occurs when an unintended electrical connection is made between two or more conductors, resulting in excessive current flow that can lead to overheating, damage to electrical components, and potentially hazardous conditions such as fire or electric shock.

Short-circuit protection systems continuously monitor the electrical circuits within the EV for signs of abnormal current flow or voltage fluctuations that may indicate a short circuit. This detection may be performed by dedicated sensors, monitoring circuits, or software algorithms embedded within the vehicle's electronic control systems.

Upon detecting a short circuit, the protection system rapidly isolates the affected circuit or component from the rest of the electrical system to prevent the spread of excessive current and mitigate potential damage. This may involve opening circuit breakers, disconnecting relays, or shutting down power to specific components or subsystems.

Sometimes, the protection system may provide visual or audible alerts to the vehicle's occupants or external monitoring systems to notify them of the detected short circuit and prompt appropriate action. This notification helps ensure that occupants know the situation and can take necessary precautions.

Once the short-circuit condition has been addressed and corrected, the protection system may automatically reset or restore power to the affected circuit or component, allowing regular operation to resume. In cases where manual intervention is required to address the short circuit, the protection system may remain deactivated until the condition has been resolved.

The actuation of short-circuit protection is essential for ensuring the safety and reliability of EVs by minimizing the risk of electrical hazards and preventing damage to critical components. By promptly detecting and isolating short circuits, these protection systems help safeguard vehicle occupants, protect valuable assets, and maintain the integrity of the vehicle's electrical infrastructure.

AER (All-Electric Range)

AER refers to the maximum distance an EV can travel on a single full charge using only its electric battery power without engaging any hybrid power sources or requiring recharging. The AER is a crucial metric for understanding the capabilities and limitations of EVs, particularly for consumers concerned with vehicle sustainability, energy efficiency, and the feasibility of using an EV for daily commutes and longer trips.

See also **Charger Anxiety**.

AID (Actuation of Isolation Detection)

AID is a safety feature in EVs designed to detect and isolate electrical faults that may occur within the high-voltage electrical system. In the event of a fault, such as a short circuit or insulation breakdown, AID systems quickly detect the fault and isolate the affected components or circuits to prevent further damage or potential safety hazards. This is achieved through sensors, circuit breakers, and control algorithms that monitor the electrical system for abnormalities and activate isolation mechanisms when necessary. By promptly isolating faults, AID systems help enhance the safety and reliability of EVs, protecting both occupants and the vehicle's electrical components from harm.

Air Clearance

Air clearance, also known as ground clearance or ride height, refers to the vertical distance between the lowest point of an EV and the ground. It is a crucial aspect of vehicle design and performance, impacting stability, aerodynamics, and off-road capability.

Ampere

An ampere, symbolized as A, is the base unit of electric current in the International System of Units (SI). It is named after André–Marie Ampère, a French mathematician and physicist considered one of the principal discoverers of electromagnetism. The ampere measures the amount of electrical charge passing a point in an electric circuit per unit of time, with one ampere equating to one coulomb of charge passing through a point in 1 sec:

$$I = \frac{P}{V \times \text{PF}}$$

where

I is the current in amperes
P is the power in watts
V is the voltage in volts (V)
PF is the power factor of the circuit (a number between 0 and 1 that represents the phase difference between the voltage and current waveforms)

Anode

Anode refers to the reducing or fuel electrode, which gives up electrons to the external circuit. Anodes, cathodes, electrolytes, and separators are critical components of lithium-ion (Li-ion) batteries commonly used in EVs and hybrid vehicles (HVs). The primary function of the anode is to facilitate the flow of ions between the anode and the cathode during the battery's charging and discharging phases. Li-ions move from the cathode to the anode when the battery is charged, where they are stored within its chemical structure. During discharge, these stored ions return to the cathode, releasing electrical energy that powers the vehicle (**Figure A.1**).

FIGURE A.1 A cutaway of the battery illustrates the anode and other components.

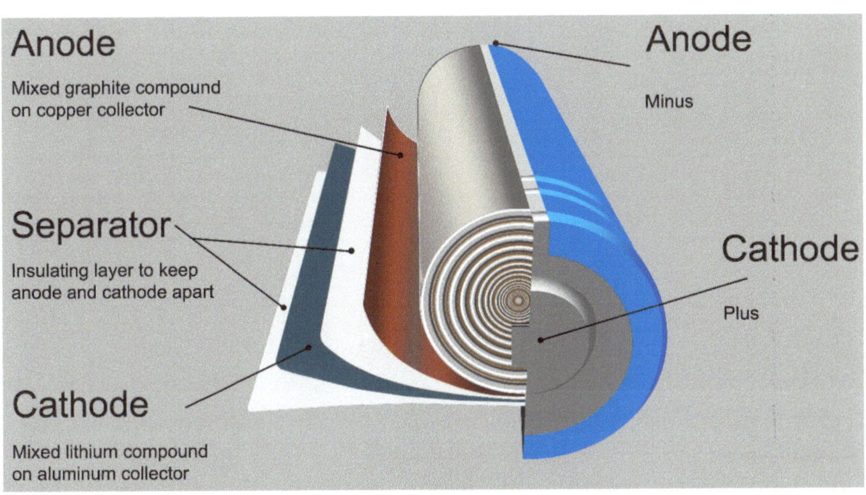

gunawand3570/Shutterstock.com.

Applied Withstand Voltage

Applied withstand voltage refers to the maximum voltage an electrical component or system within an EV can safely withstand without experiencing damage or failure. This specification is crucial for ensuring the safety and reliability of EV components, particularly in high-voltage systems such as batteries, power electronics, and electrical wiring. Applied withstand voltage is typically expressed in volts (V) and represents the maximum voltage level the component or system can endure under normal operating conditions and during transient or faulty conditions. Compliance with applied withstand voltage requirements is essential to prevent electrical arcing, insulation breakdown, and other hazards that could pose risks to vehicle occupants and ensure EV systems' overall performance and longevity.

APU (Auxiliary Power Unit)

A device used to generate propulsion power but without a direct mechanical connection to the drive wheels. Some examples of auxiliary power units are ICEs connected solely to an electric generator, an engine connected solely to a hydraulic pump, or fuel cells.

For additional information, see **J1715** [3].

Arc Flash Hazard Calculation

Arc flash hazard calculation determines the potential arc flash energy and associated risks in an electrical system or equipment. An arc flash is a dangerous electrical discharge when an electrical current passes through the air between conductors or between a conductor and the ground. This discharge can produce intense heat, light, and pressure, leading to severe burns, injuries, and damage to equipment.

A generic form of the arc flash incident energy equation is expressed as follows:

$$E = \frac{\left(4.184 \times K \times G \times (\tau) \times \left(I_f\right)^2\right)}{D^2}$$

where

E is the incident energy in joules per centimeter squared (J/cm^2)
K is a constant derived from system voltage and configuration
G is a grounding factor
τ is the arc duration in seconds
I_f is the fault current in amperes
D^2 is the distance from the arc flash to the person or object in centimeters

Arc flash hazard calculations involve analyzing various factors, including the available fault current, the time it takes for protective devices to operate, the arc flash boundary, and the incident energy levels at different distances from the arc. These calculations help determine the level of personal protective equipment (PPE) required for workers working on or near energized electrical equipment to minimize the risk of injury during an arc flash.

Authentication Control

Authentication control in the context of EVs refers to the security measures implemented to verify the identity and authorization of users accessing the vehicle's systems or features. These control mechanisms help prevent unauthorized access, theft, or misuse of the vehicle and its functionalities. Authentication control typically involves using authentication methods such as passwords, biometric recognition (e.g., fingerprint or facial recognition), smart cards, or key fobs to authenticate the user's identity before granting access to the vehicle or its connected services. By enforcing authentication control, EV manufacturers and service providers ensure the security and integrity of vehicle systems, protect user privacy, and mitigate the risk of unauthorized access or malicious activities.

Automatic Hazardous Voltage Disconnects

Automatic hazardous voltage disconnects are safety mechanisms designed to automatically disconnect high-voltage circuits in EVs in the event of a collision or other dangerous situations. These disconnects are implemented as part of the vehicle's high-voltage electrical system to mitigate the risk of electric shock and fire in the event of a crash or malfunction.

The automatic hazardous voltage disconnects function by sensing abnormal conditions such as sudden changes in vehicle acceleration, impact forces, or electrical system abnormalities. Upon detecting such situations, the disconnects rapidly interrupt the flow of high-voltage electricity, isolating the battery pack and electrical components from the rest of the vehicle's systems.

By automatically disconnecting high-voltage circuits during hazardous events, these safety mechanisms help minimize the risk of electric shock to occupants and first responders and reduce the likelihood of post-collision fires caused by electrical faults. Automatic hazardous voltage disconnects are essential safety features in EVs, protecting occupants, emergency responders, and bystanders during accidents or emergencies.

Auto-Spectral Density

Auto-spectral density, also known as auto-power spectral density (PSD), is a statistical measure used in signal processing and vibration analysis to describe how a signal's power is distributed across different frequencies. In EVs, auto-spectral density plays a crucial role in analyzing and optimizing the performance and efficiency of various components, such as electric motors, batteries, and power electronics systems.

Auto-spectral density is used to identify and analyze electric motors' vibration and noise characteristics. Engineers can design measures to reduce noise, improve comfort, and enhance motor efficiency by understanding the frequency distribution of the motor's vibrations.

For battery systems, auto-spectral density can help diagnose issues related to battery cell vibrations, which might indicate structural weaknesses or potential failures. Monitoring the spectral density of vibrations can lead to early detection of problems, extending battery life and ensuring safety.

Auto-spectral density analysis helps identify and mitigate electromagnetic interference (EMI) in power electronics, including inverters and converters [6]. By analyzing the frequency distribution of EMI, engineers can implement strategies to reduce interference, improving the reliability and performance of electronic components.

Auto-spectral density is also instrumental in monitoring EVs' structural health. Analyzing the frequency content of vibrations throughout the vehicle can help identify areas of stress or potential failure, leading to more targeted maintenance and longer vehicle lifespans.

Availability Control

Availability control in connected vehicles refers to a system or feature designed to ensure that essential functions and services within the vehicle remain accessible and operational, even in challenging or adverse conditions. It involves monitoring, managing, and optimizing the availability of critical vehicle systems, including communication, navigation, safety, and entertainment features, to enhance the overall reliability and usability of the vehicle.

AVAS (Acoustic Vehicle Alerting System)

An AVAS is a safety feature installed in EVs to generate audible sounds that alert pedestrians and other road users to the vehicle's presence. Since EVs operate silently at low speeds, they may pose a risk to pedestrians, especially those who are visually impaired, as they may not hear the vehicle approaching.

The AVAS emits sound patterns or tones that mimic the sound of traditional ICE vehicles, providing an auditory cue to alert pedestrians to the vehicle's presence. The AVAS typically activates at low speeds, such as during start-up, reversing, or when traveling at low speeds in urban areas, and deactivates at higher speeds when tire and wind noise become sufficient for pedestrian awareness. The primary purpose of AVAS is to enhance pedestrian safety and mitigate the risk of accidents involving EVs.

REFERENCES

[1]. Society of Automotive Engineers, "J3068 Identification of Vehicles and Supply Equipment for Conductive AC Charging," SAE Publishing, Warrendale, 2023.

[2]. Society of Automotive Engineers, "J1772 SAE Electric Vehicle and Plug-In Hybrid Electric Vehicle Conductive Charge Coupler," SAE Publishing, Warrendale, 2024.

[3]. Society of Automotive Engineers, "J1715 Hybrid Electric Vehicle (HEV) and Electric Vehicle (EV) Terminology," SAE Publishing, Warrendale, 2022.

[4]. IEEE Standards Association, "IEEE Std C95.-2019 IEEE Standard for Safety Levels with Respect to Human Exposure to Electric, Magnetic, and Electromagnetic Fields, 0 Hz to 300 GHz," IEEE, 2019.

[5]. Society of Automotive Engineers, "J553 Circuit Breakers," SAE Publishing, Warrendale, 2022.

[6]. Society of Automotive Engineers, "J1113-12 Electrical Interference by Conduction and Coupling - Capacitive and Inductive Coupling via Lines Other than Supply Lines," SAE Publishing, Warrendale, 2022.

B

Back EMF (Electromotive Force)

Back EMF [1, 2] is a phenomenon observed in electric motors, including those used in EVs. It refers to the voltage generated in an electric motor's windings when rotating but not actively receiving electrical power. When an electric motor is in motion, the rotation of the armature induces a voltage in the motor's windings, opposing the applied voltage that drives current through the motor. This induced voltage, known as Back EMF, is proportional to the motor's rotational speed and is an essential factor in controlling the motor's speed and torque characteristics. Back EMF plays a crucial role in EVs in motor control algorithms, allowing for efficient operation, precise speed control, and regenerative braking capabilities. Understanding Back EMF is essential for optimizing the performance and efficiency of EV propulsion systems.

Battery

In the context of EVs, a battery is the energy storage device responsible for supplying electrical power to the vehicle's electric motor(s) for propulsion. EV batteries typically consist of multiple individual cells connected in series and parallel to form a battery pack. These cells use chemical reactions to store and release electrical energy, allowing the EV to operate without relying on ICEs.

Modern EV batteries predominantly utilize Li-ion chemistry due to their high energy density, lightweight, and long cycle life. The capacity of an EV battery, measured in kilowatt-hours (kWh), determines the amount of energy it can store and, subsequently, the vehicle's driving range on a single charge.

Battery Capacity

Battery capacity refers to the amount of electrical energy a battery can store and deliver to power an EV. It is typically measured in kWh and represents the total energy capacity available. A higher battery capacity allows an EV to travel greater distances on a single charge, providing increased range and driving flexibility. Battery capacity is a critical factor in determining the performance, range, and usability of EVs, and it directly influences factors such as driving range, charging frequency, and overall driving experience. Understanding the battery capacity of an EV is essential for evaluating its suitability for specific driving needs and comparing different EV models.

$$\text{Capacity}\left(\text{Ah}\right)=\frac{\text{Energy}\left(\text{Wh}\right)}{\text{Voltage}\left(\text{V}\right)}$$

where

Capacity is the battery's capacity in ampere-hours (Ah)
Energy is the total energy stored in the battery in watt-hours (Wh)
Voltage is the battery's nominal voltage in volts (V)

Another way to express battery capacity is by using the current and time, which can be represented as

$$\text{Capacity}\left(\text{Ah}\right)=\text{Current}\left(\text{A}\right)\times\text{Time}\left(\text{h}\right)$$

where

Current is the constant discharge current in amperes (A) that the battery can provide
Time is the duration in hours (h) for which the battery can provide this current

Battery Degradation

Battery degradation refers to the gradual loss of capacity and performance over time in an EV battery pack. This degradation is a natural phenomenon resulting from various chemical and physical processes within the battery cells during charging and discharging cycles (Figure B.1). As the battery undergoes repeated charge and discharge cycles, its ability to store and deliver electrical energy gradually diminishes, reducing driving range and overall battery performance [3].

FIGURE B.1 Li-ion cell degradation mechanisms [5].

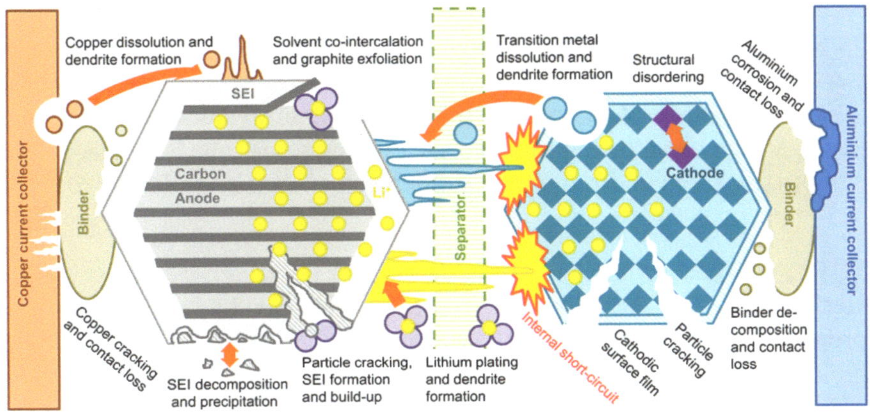

Battery degradation in EVs due to aging is a complex process influenced by many factors, and there is not a single, universally accepted equation to describe it [4]. However, battery degradation is commonly modeled using factors like cycle count, depth of discharge (DoD), temperature, and current. A simplified equation that reflects some of these factors is

$$C_{remaining} = C_{initial} \times (1 - k \times \eta)$$

where

$C_{remaining}$ is the remaining battery capacity after n cycles
$C_{initial}$ is the initial battery capacity
k is the rate of capacity loss per cycle
η is the number of charge–discharge cycles

A few things contribute to battery degradation. Each charging and discharging cycle causes chemical changes within the battery cells, leading to the breakdown of electrode materials and loss of active material. Over time, this cycle aging process decreases battery capacity and performance.

Even when not used, batteries undergo degradation over time due to chemical reactions within the cells. This calendar aging is influenced by temperature, SOC, and storage conditions, which can accelerate the degradation process.

Exposure to high temperatures, whether during operation or storage, accelerates battery degradation by promoting chemical reactions that degrade electrode materials and increase internal resistance. High temperatures also accelerate electrolyte decomposition, reducing the battery's ability to store energy effectively. Operating the battery consistently at high states of charge or deeply discharging it can accelerate degradation by increasing the likelihood of side reactions and material degradation within the battery cells. Rapid charging, especially at high charging rates, generates heat and stress within the battery cells, which can accelerate the degradation of electrode materials and decrease battery lifespan.

Variations in the SOC or performance among individual cells within the battery pack can lead to cell imbalances, causing some cells to degrade more rapidly than others and reducing overall battery capacity and performance.

Battery Edge

Battery edge refers to the state or condition of an EV battery pack when it approaches its upper or lower operational limits. When the battery is at the edge, it is either nearing its maximum SOC or minimum SOC, depending on whether it is being charged or discharged. Operating the battery at its edge for prolonged periods can affect battery health, performance, and longevity.

For example, when the battery is nearing its upper SOC limit during charging, it may operate at a high voltage, which can increase the risk of overcharging and thermal runaway if not correctly managed. Conversely, when the battery approaches its lower SOC limit during discharge, it may be at risk of deep discharge, leading to capacity degradation and reduced battery lifespan.

Battery management systems (BMSs) in EVs are crucial in monitoring and controlling the battery's SOC to prevent it from reaching the edge under normal operating conditions. The individual cells of a pack may experience edge conditions in a non-uniform manner because of cell-to-cell variations. The "edge" of a pack is thus defined by its least capable cell(s) and must be managed accordingly. By carefully managing charging and discharging processes, the BMS helps optimize battery performance, efficiency, and longevity while ensuring the safe and reliable operation of the EV.

Battery Heating System

A battery heating system is a component of an EV designed to regulate the temperature of the traction battery pack, particularly in cold weather conditions. The system utilizes various heating elements, such as electric resistive heaters or liquid-based heating systems, to raise the battery pack's temperature to an optimal operating range. By maintaining the battery temperature within the recommended range, the heating system helps to improve battery performance, efficiency, and longevity, especially during cold weather when low temperatures can adversely affect battery performance and range. Additionally, a battery heating system may facilitate faster charging by preheating the battery pack before charging, reducing charging times and improving overall vehicle usability in cold climates. Overall, the battery heating system plays a crucial role in ensuring reliable operation and optimal performance of EVs in diverse weather conditions.

Battery Module

A battery module is a self-contained unit within an EV's battery pack, designed to store and provide electrical energy to power the vehicle's electric drivetrain (Figure B.2). It consists of multiple interconnected battery cells, typically Li-ion cells, enclosed within a protective casing along with management systems for monitoring and controlling various parameters such as temperature, voltage, and SOC.

FIGURE B.2 A cross section of the battery module showing constituent parts [6].

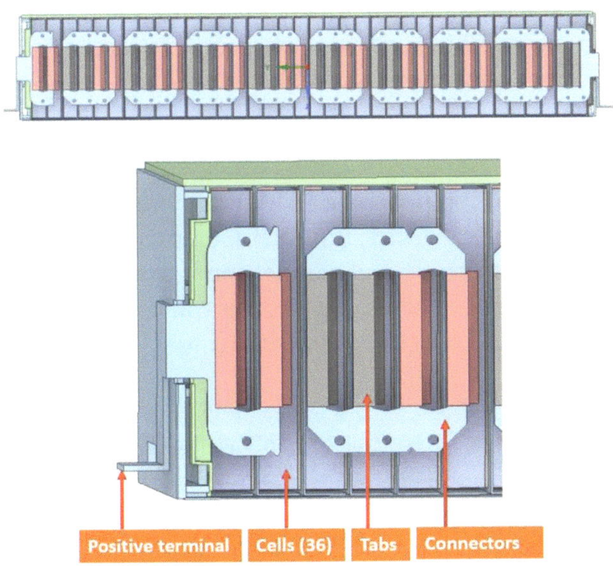

Battery modules are the building blocks of the larger battery pack in EVs. They store electrical energy generated by charging sources, such as grid electricity or regenerative braking, and deliver power to the vehicle's electric motor during operation. Each module has a BMS that regulates the charging and discharging processes, ensuring the battery pack's safety, longevity, and optimal performance.

A battery module typically comprises several cells connected in series and parallel configurations to achieve the desired voltage and capacity specifications. These cells are often arranged compactly and efficiently within the module casing, designed to provide mechanical support, thermal management, and protection against environmental factors and potential hazards.

Battery Overcharge

Battery overcharge refers to the condition in which the battery of an EV receives more electrical energy than it can safely store or manage. This excess charging can lead to various adverse effects, including thermal runaway, electrolyte decomposition, and degradation of battery components. Overcharging typically occurs when the charging process continues beyond the battery's maximum recommended SOC, causing the voltage and temperature of the battery cells to exceed safe limits. Overcharging can result from malfunctioning charging equipment, incorrect charging settings, or software errors in the BMS. Over time, repeated overcharging can lead to reduced battery capacity, decreased performance, and potential safety hazards, underscoring the importance of implementing proper charging protocols and safeguards to prevent overcharge incidents in EVs.

Battery Pack

A battery pack is a crucial component of an EV that stores and supplies electrical energy to power the vehicle's electric motor(s). Comprising multiple individual battery cells connected in series and parallel configurations, a battery pack is designed to provide the necessary voltage, current, and energy capacity to meet the EV's performance and range requirements (Figure B.3).

FIGURE B.3 The battery pack is a collection of battery cells configured to power the vehicles.

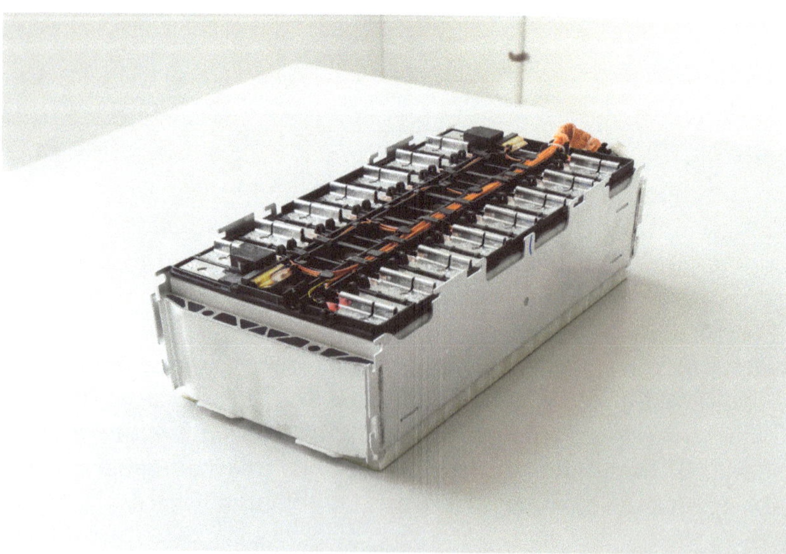

Smile Fight/Shutterstock.com.

The battery pack serves as the primary energy storage system (ESS) of the EV, storing electrical energy obtained from external charging sources and releasing it as needed to propel the vehicle. It typically consists of Li-ion or other advanced battery chemistries optimized for high energy density, efficiency, and longevity.

Critical components of a battery pack include cell modules, BMS, thermal management systems, and structural enclosures. The BMS monitors and manages individual battery cells' SOC, temperature, and voltage to ensure safe and optimal operation. At the same time, the thermal management system regulates battery temperature to prevent extremes of heat or cold maintaining a thermal operational envelope.

Battery packs vary in size, capacity, and configuration depending on the specific requirements of the EV platform, such as vehicle size, weight, and intended use. Advanced engineering and manufacturing techniques are employed to design and produce battery packs that deliver high performance, reliability, and safety while meeting the automotive industry's stringent requirements.

The battery pack is critical in determining an EV's overall performance, range, and driving experience. As battery technology advances, battery packs are evolving to offer higher energy density, faster charging capabilities, and increased durability, driving the widespread adoption of EVs as a sustainable transportation solution.

Battery Pack System

The battery pack system in an EV refers to the integrated assembly of individual battery cells, modules, and associated components designed to store and deliver electrical energy to power the vehicle's electric motor(s). It serves as the primary energy storage unit in the EV, providing the necessary power to propel the vehicle and operate auxiliary systems (Figure B.4).

Battery cells are the basic building blocks of the battery pack. They are electrochemical devices that store electrical energy through chemical reactions. They are typically arranged in series and parallel configurations to achieve the desired voltage and capacity for the EV.

Battery modules are groups of interconnected battery cells housed within protective enclosures. They provide structural support and thermal management for the individual cells while facilitating electrical connections and balancing between cells.

Battery pack systems are equipped with cooling and thermal management systems to maintain optimal operating temperatures and prevent overheating. These may include liquid or air-cooled heat exchangers, coolant circulation systems, and temperature sensors to regulate battery temperature and dissipate excess heat generated during charging and discharging.

FIGURE B.4 The battery pack system is an integral part of the vehicle propulsion system.

PP77LSK/Shutterstock.com.

The battery pack is housed in a structural enclosure or casing to protect the cells and components from external impacts, vibrations, and environmental hazards. The enclosure also provides mechanical support and ensures proper integration within the vehicle chassis.

Battery pack systems incorporate various safety features to prevent and mitigate risks associated with battery operation, such as thermal runaway, short circuits, and electrical faults. These may include safety disconnects, fuse protection, flame retardant materials, and redundant safety systems to ensure safe operation under normal and faulty conditions.

Battery Percent Desired

Battery percent desired refers to the preferred SOC or battery level that an EV owner or driver aims to achieve before starting or continuing a journey. It is typically expressed as a percentage of the total battery capacity available. Setting a specific battery percent desired helps EV drivers plan their trips effectively by ensuring they have sufficient charge to reach their destination comfortably while optimizing charging opportunities along the way. This value may vary depending on driving distance, route conditions, charging infrastructure availability, and personal preferences. Maintaining an adequate battery percentage helps mitigate range anxiety and ensures a smooth and stress-free driving experience in EVs.

Battery Power Converter

A battery power converter, also known as a DC–DC converter, is a component in EVs responsible for managing the electrical power flow between the high-voltage battery pack and various vehicle systems. The primary function of the battery power converter is to convert the high-voltage DC output from the traction battery into lower voltages suitable for powering auxiliary systems, such as the onboard electronics, lighting, HVAC, and charging the low-voltage auxiliary battery.

The battery power converter ensures efficient energy utilization from the main traction battery, maintaining optimal voltage levels and providing a stable power supply to the vehicle's electrical components. Additionally, it may incorporate safety features, such as overcurrent protection and overvoltage protection to safeguard the battery pack and vehicle electronics from potential damage or malfunction.

In hybrid EVs (HEVs) and plug-in HEVs (PHEVs), the battery power converter facilitates bidirectional power flow, transferring energy between the traction battery and the vehicle's electrical system. It also enables regenerative braking, where kinetic energy is recovered and stored in the battery pack during deceleration.

Battery Simulator

A battery simulator is a sophisticated electronic device to develop and test EVs and their associated components, particularly BMS. It mimics the behavior of an EV's battery pack, allowing engineers to simulate various operating conditions and scenarios without needing a physical battery.

Battery SOC (State of Charge)

Battery SOC refers to the current level of electrical energy stored in an EV battery pack, expressed as a percentage of the total capacity (Figure B.5). It indicates how much energy is available to power the vehicle's electric motor and determines the driving range the EV can achieve on a single charge. A higher SOC corresponds to a more significant amount of stored energy and a more extended driving range. In comparison, a lower SOC indicates that the battery has less available energy and may require recharging soon. Monitoring the battery's SOC is essential for EV drivers to plan their journeys effectively and ensure they have sufficient range to reach their destination without running out of battery power.

FIGURE B.5 Vehicles come with a SOC display providing information on range and often percent charge remaining.

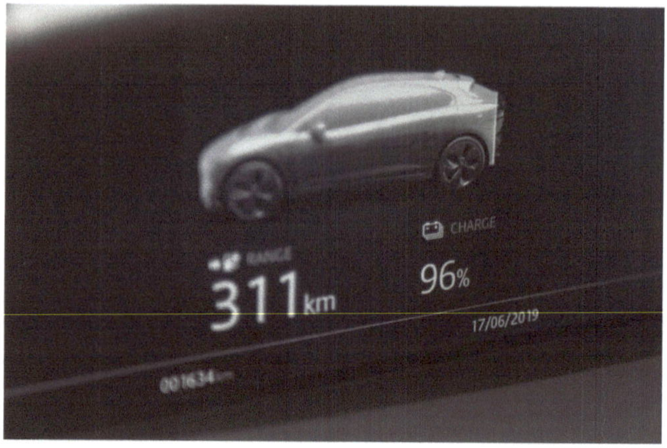

supergenijalac/Shutterstock.com.

BEV (Battery Electric Vehicle)

A BEV is an EV powered solely by electricity stored in an onboard battery pack. Unlike HVs, which combine an ICE with an electric motor and battery, BEVs rely exclusively on electric propulsion, with no tailpipe emissions. BEVs use rechargeable Li-ion batteries or other advanced battery technologies to store electrical energy, which powers one or more electric motors to drive the vehicle. BEVs offer zero-emission driving and are increasingly popular as sustainable alternatives to traditional gasoline-powered vehicles, offering benefits such as reduced air pollution, lower operating costs, and energy independence. Examples of BEVs include models like the Tesla Model S, Nissan Leaf, and Chevrolet Bolt EV.

BEV Energy Consumption and Range Test Procedure

The BEV Energy Consumption and Range Test Procedure is a standardized methodology for measuring and evaluating BEVs' energy consumption and driving range [7]. This test procedure is developed by regulatory agencies, industry organizations, and vehicle manufacturers, and it provides a consistent and repeatable framework for assessing BEVs' energy efficiency and range performance under controlled conditions (Table B.1).

TABLE B.1 An example of the parameters to ascertain the battery consumption and range [7].

Item	Parameter/Measurement	Description	Units
1	Battery ampere-hour capacity	Net DC ampere-hours discharged during test	DC A·h
2	Ambient test temperature	—	°C (°F)
3	Time, soak start	Pre-test soak time	—
4	Time, soak end	Pre-test soak time	—
5	Time, test start	—	—
6	Distance driven (total)	—	km (mi)
7	Distance driven per phase ($D_{[cycle]\_i}$)	—	km (mi)
8	Discharge energy[1] (Edc_{total}, $Edc_{totalN7C}$[4])	Net DC watt-hours discharged during test	DC W·h
9	Discharge energy ($Edc_{discharge}$[5], $Edc_{dischargeN7C}$[4])	Net DC watt-hours discharged during DC discharge test procedure	DC W·h
10	Phase and cycle discharge energy ($Edc_{[phase]\_i}$, $Edc_{[cycle]\_i}$)	Net DC watt-hours discharged per phase and/or per cycle	DC W·h
11	Time, end of test	—	—
12	Vehicle charging mode	If multiple settings available	—
13	Time, start of charge	Time connected to EVSE	—
14	Power outlet voltage	Nominal AC voltage at power outlet	AC volts
15	AC recharge voltage[2]	Real-time RMS voltage measured at power outlet (one sample per minute minimum)	AC volts
16	AC recharge energy (Eac_{post}, FRE_{post} Eac_{pre}[3], FRE_{pre}[3])	Total AC energy from power outlet	AC W·h
17	Time, end of vehicle soak period	Total time vehicle is soaking	—
18	Time, end of recharging period	Time full charge (FC) is achieved (may include thermal conditioning time)	—

[1] Optional for SCT
[2] Recommended
[3] Thermal conditioning only
[4] Five-cycle procedure only
[5] SMCT only

Bi-Directional Inverter

A bi-directional inverter, also known as a bidirectional converter or power electronics unit, is a critical component in EVs capable of both converting DC power from the vehicle's battery into AC for propulsion, as well as converting AC power back into DC for energy storage or converting back to AC for grid interaction purposes (Figure B.6) [3, 8]. Unlike traditional inverters that only convert DC to AC for propulsion, a bi-directional inverter enables bidirectional power flow, allowing the vehicle to operate in both charging and discharging modes.

FIGURE B.6 This is a system concept for a plug-in EV (PEV) with an onboard inverter system interconnecting to the electric power system by way of a conductive coupling to EV supply equipment (EVSE) [8].

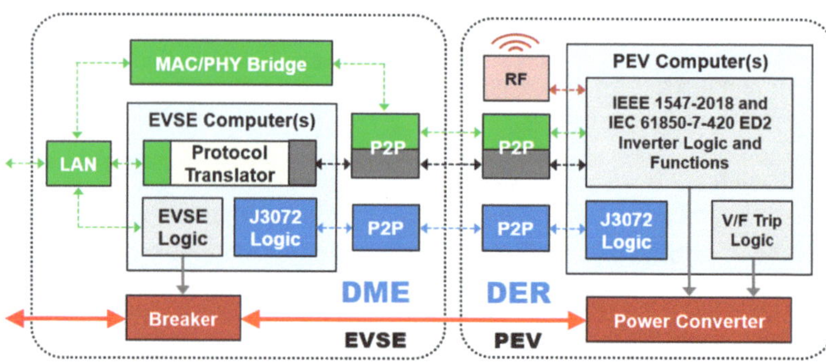

In charging mode, the bi-directional inverter converts AC power from an external power source, such as a charging station or grid, into DC power to recharge the vehicle's battery pack. This allows EVs to replenish their battery capacity from external power sources, such as residential or commercial charging stations.

In discharging mode, the bidirectional inverter converts DC power from the vehicle's battery into AC power for propulsion, enabling the electric motor to drive the vehicle's wheels. This is the inverter's primary function during normal driving operations.

Bi-directional inverters also enable vehicle-to-grid (V2G) operation, allowing EVs to discharge power back into the grid during periods of high demand or to provide grid stabilization services. This capability enables EVs to act as mobile energy storage units, contributing to grid reliability and supporting renewable energy integration.

Some bi-directional inverters support vehicle-to-home (V2H) operation, allowing EVs to power residential or commercial buildings during power outages or peak demand periods. V2H systems enhance energy resilience and reduce reliance on traditional grid infrastructure by using the vehicle's battery as a backup power source.

BMS (Battery Management System)

A BMS is a critical component in EVs responsible for monitoring, controlling, and optimizing the battery pack's performance, safety, and lifespan (Figure B.7). The BMS functions as the "brain" of the battery system, continuously monitoring various parameters such as voltage, current, temperature, and SOC of individual battery cells or modules.

FIGURE B.7 The BMS and interactions with vehicle subsystems.

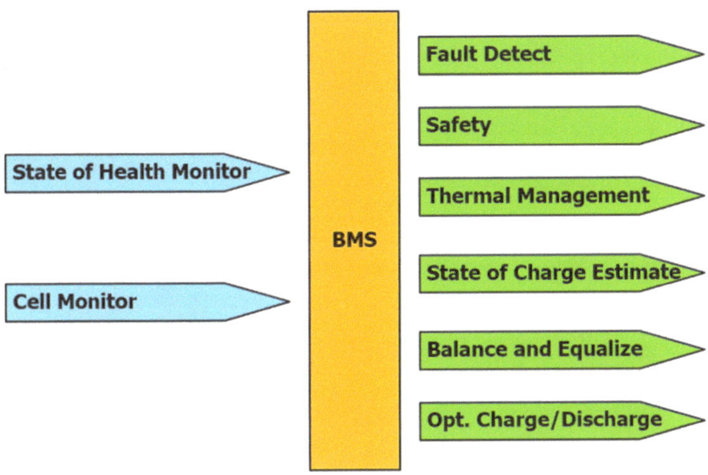

The BMS monitors each battery cell or module's voltage, current, and temperature to ensure balanced charging and discharging and prevent over-charging, overdischarging, or overheating individual cells. By analyzing voltage and current measurements, the BMS estimates the battery pack's SOC, providing accurate information to the vehicle's onboard systems and driver displays.

The BMS tracks the health and degradation of the battery pack over time, identifying any abnormalities or degradation trends that may affect performance or lifespan. The BMS controls thermal management systems, such as cooling or heating systems, to maintain optimal operating temperatures for the battery cells, ensuring safety and maximizing performance and longevity.

In multicell battery packs, the BMS manages cell balancing and equalization to ensure all cells have a uniform charge state, preventing overcharging individual cells and maximizing overall battery capacity and lifespan. The BMS includes safety features such as overvoltage protection, undervoltage protection, overcurrent protection, and short-circuit protection to prevent hazardous conditions and ensure the safe operation of the battery pack. The BMS monitors for faults or abnormalities in the battery system and provides diagnostics and fault codes to alert the vehicle's onboard systems and service technicians of any issues that require attention. Based on battery condition, temperature, and other factors, the BMS controls charging and discharging processes to optimize efficiency, minimize degradation, and extend battery lifespan.

Boundary Protection

Boundary protection refers to safety measures and systems implemented in EVs to safeguard critical components, such as the battery pack, powertrain, and electronic systems, from potential hazards or external factors that may compromise their integrity or performance. These protective measures establish physical and operational boundaries to prevent overloading, overheating, short circuits, water ingress, and other adverse conditions that could damage or malfunction.

Critical components of boundary protection systems in EVs may include systems that monitor and limit the flow of electrical current to prevent overheating and damage to electrical components, such as fuses, circuit breakers, and current-limiting devices. Thermal management systems regulate the temperature of critical components, such as the battery pack and power electronics, to prevent overheating and ensure optimal performance and longevity. This may involve active cooling, passive cooling, or passive measures to protect electrical and electronic components from water ingress, moisture, and corrosion, including waterproof seals, gaskets, and encapsulation of vulnerable components. Systems that regulate the voltage supplied to electrical components ensure they operate within safe limits and prevent damage from overvoltage conditions. It will also include monitoring systems that detect abnormalities, faults, or malfunctions in electrical circuits and components and isolate affected areas to prevent further damage or safety hazards; shielding measures to protect sensitive electronic components from EMI generated by external sources or nearby electrical systems; and reducing the risk of performance degradation or malfunction. Protective enclosures, barriers, and shielding help to prevent physical damage to critical components from external impacts, debris, or environmental hazards.

Brake Recuperation

Brake recuperation, also known as regenerative braking or regen braking, is a technology employed in EVs to recover energy during braking and deceleration [9, 10]. When the driver applies the brakes or reduces the vehicle's speed, the electric motor(s) in the EV act as generators, converting kinetic energy into electrical energy (Figure B.8). This electrical energy is then stored in the vehicle's battery pack for later use, such as powering the vehicle's electric drivetrain or accessories. Brake recuperation enhances the efficiency of EVs by harnessing energy that would otherwise be dissipated as heat during traditional friction braking, extending the vehicle's driving range and reducing energy consumption. Additionally, regenerative braking systems can provide drivers with a smoother, more controlled braking experience, enhancing vehicle safety and comfort.

See also **Regenerative Braking**.

FIGURE B.8 Experimental battery current measurements at different speed/torque operating points of the permanent magnet synchronous motor (PMSM) show the low-speed boundary (black line) and motor control center (MCC) (dashed red line) [9].

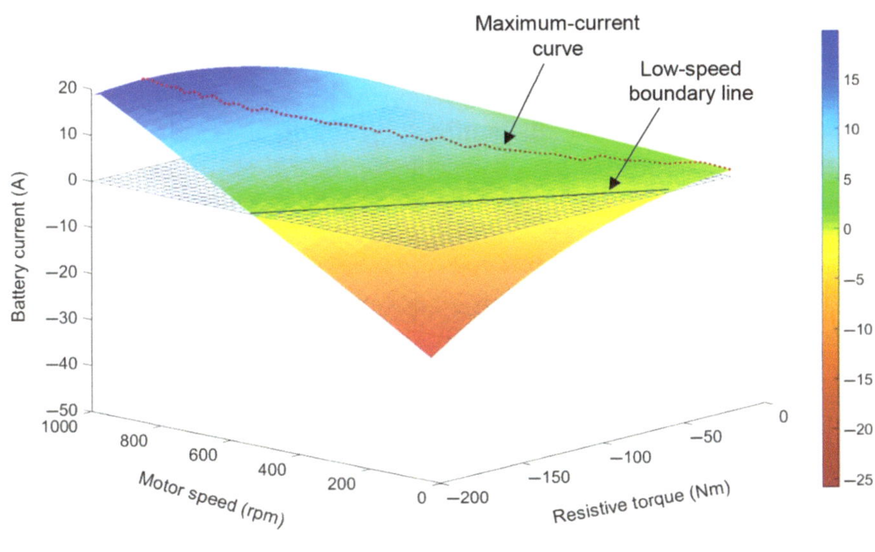

Branch Circuit Breaker Rating

The branch circuit breaker rating refers to the maximum electrical current capacity that a circuit breaker can handle in a specific branch of an electrical distribution system without tripping. In the context of EVs, this term is particularly relevant for the infrastructure that supplies electric power for EV charging stations.

For EV charging, the branch circuit breaker rating is crucial because it determines the maximum current that can be safely supplied to an EV charger without exceeding the circuit breaker's capacity. This rating is a safety mechanism that prevents overloading of the electrical circuit, which can lead to overheating, fire hazards, and damage to the electrical infrastructure and the EV charger.

Broadband PLC (Power Line Communication) for PEV

Broadband PLC for PEVs is a technology that enables data transmission over electrical power lines within the vehicle charging infrastructure [11]. This form of communication uses the existing electrical wiring to send and receive data, facilitating various functionalities such as charging management, V2G communication, and the integration of EVs into intelligent grid applications.

Broadband PLC allows communication between the EV and the charging station, enabling features like authentication, billing, and real-time monitoring of charging status. It also helps optimize the charging process based on the grid's capacity and the vehicle's requirements.

It supports V2G services, where EVs can return electricity to the grid, especially during peak demand times. This requires precise communication to manage the flow of electricity back to the grid efficiently.

By facilitating seamless communication between EVs and the smart grid, broadband PLC helps manage demand response (DR) and load balancing and enhances the electrical grid's overall efficiency.

Broadband PLC technology for EVs operates over a wide range of frequencies, typically from a few kilohertz (kHz) to several megahertz (MHz), allowing for high-speed data transmission. This communication technology can coexist with the AC power flow, making it a cost-effective data exchange solution without additional wiring (Figure B.9).

FIGURE B.9 A depiction of an implementation of parallel injection [11].

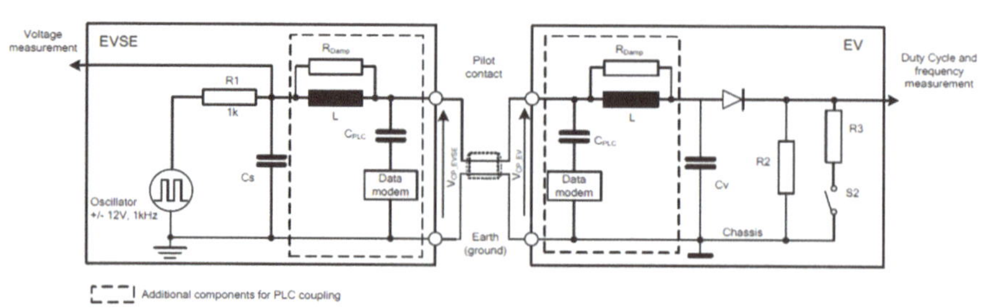

BSS (Battery Swapping Station)

A BSS is a specialized facility designed to rapidly exchange depleted batteries in EVs with fully charged ones (Figure B.10). Unlike traditional charging stations where EVs are connected to a charger and charged over a period, BSSs allow for the quick replacement of batteries, significantly reducing the time required for recharging. At a BSS, EV owners drive their vehicles onto a platform or docking station, where automated machinery or personnel removes the depleted battery pack and replaces it with a fully charged one. This process typically takes only a few minutes, providing EV owners with a convenient and efficient way to extend their driving range without waiting for a battery to recharge. BSSs can help address challenges associated with long charging times and limited charging infrastructure, particularly in regions where EV adoption is increasing. They offer a viable alternative to traditional charging methods, offering increased convenience and flexibility for EV owners.

FIGURE B.10 An example of a BSS.

khiara aqila arisanti/Shutterstock.com.

REFERENCES

[1]. Society of Automotive Engineers, "J1113-21 Electromagnetic Compatibility Measurement Procedure for Vehicle Components - Part 21: Immunity to Electromagnetic Fields, 30 MHz to 18 GHz, Absorber-Lined Chamber," SAE Publishing, Warrendale, 2013.

[2]. Society of Automotive Engineers, "J1113-12 Electrical Interference by Conduction and Coupling - Capacitive and Inductive Coupling via Lines Other than Supply Lines," SAE Publishing, Warrendale, 2022.

[3]. Society of Automotive Engineers, "J1715 Hybrid Electric Vehicle (HEV) and Electric Vehicle (EV) Terminology," SAE Publishing, Warrendale, 2022.

[4]. Society of Automotive Engineers, "J2288 Life Cycle Testing of Electric Vehicle Battery Modules," SAE Publishing, Warrendale, 2020.

[5]. C. Gera and S. Holavanahalli, *Optimizing the Dispense Time of a Battery from the Swappable Charging Station*, Warrendale: SAE Publishing, 2024.

[6]. P. Ding, W. Jiang, A. Majumdar, P. Pawar, X. Hu and A. Wakale, "Thermal Reduced Order Modeling for System Analysis of EV Battery," *SAE Int. J. Adv. & Curr. Prac. in Mobility*, vol. 6, no. 1, pp. 377-384, 2023, https://doi.org/10.4271/2023-01-0931.

[7]. Society of Automotive Engineers, "J1634 Battery Electric Vehicle Energy Consumption and Range Test Procedure," SAE Publishing, Warrendale, 2021.

[8]. Society of Automotive Engineers, "J3072 Interconnection Requirements for Onboard, Grid Support Inverter Systems," SAE Publishing, Warrendale, 2021.

[9]. S. Heydari, P. Fajri, N. Lotfi and M. Rasheduzzaman, "Maximum-Current Curve Operation of Electric Vehicles for Improved Energy Recuperation during Regenerative Braking," *SAE Int. J. Elec. Veh.*, vol. 12, no. 2, pp. 145-155, 2023, https://doi.org/10.4271/14-12-02-0007.

[10]. Y. Yao, Y. Zhao and M. Yamazaki, "Integrated Regenerative Braking System and Anti-Lock Braking System for Hybrid Electric Vehicles & Battery Electric Vehicles," *SAE Int. J. Adv. & Curr. Prac. in Mobility*, vol. 2, no. 3, pp. 1592-1601, 2020.

[11]. Society of Automotive Engineers, "J2931-4 Broadband PLC Communication for Plug-In Electric Vehicles," SAE Publishing, Warrendale, 2023.

C

CaaS (Charging as a Service)

CaaS refers to a business model in which EV owners can access charging infrastructure and services on a pay-per-use or subscription basis rather than owning or maintaining charging equipment. Under this model, third-party providers or operators install, manage, and maintain EV charging stations at various locations, such as public parking facilities, commercial establishments, and residential complexes. EV owners can access these charging stations conveniently through mobile apps, radiofrequency identification (RFID) cards, or other payment methods, paying for the energy consumed during charging sessions. CaaS offering may include various features and benefits, such as access to a network of charging stations, real-time monitoring and remote management capabilities, billing and payment processing, and customer support services. This model aims to address barriers to EV adoption, such as the availability and accessibility of charging infrastructure, by providing convenient and reliable charging solutions for EV owners while enabling monetization opportunities for charging service providers [1].

Captive Charging Station

A captive charging station is a dedicated EV charging facility installed and managed by a specific entity, typically for private or restricted use. Unlike public charging stations that are accessible to any EV owner, captive charging stations are often installed in locations such as residential complexes, corporate campuses, fleet depots, or government facilities, where access is limited to authorized users or tenants [2]. Captive charging stations may offer various charging levels, including Level 1, Level 2, or DC fast charging, to accommodate different types of EVs and charging needs [3]. These stations may have access controls, billing systems, and network connectivity to monitor usage, manage access permissions, and facilitate billing for charging services. Captive charging stations play a crucial role in supporting the adoption of EVs by providing convenient and accessible charging solutions for specific user groups or organizations.

Cathode

In EVs and battery technology, the cathode is a crucial component of the Li-ion battery pack. It is one of the two electrodes within the battery, the other being the anode. The cathode serves as the electrode where positively charged ions, such as lithium ions, migrate during the charging process.

The cathode is typically composed of various materials, with lithium cobalt oxide ($LiCoO_2$), lithium iron phosphate ($LiFePO_4$), and lithium manganese oxide ($LiMn_2O_4$) being common choices. These materials facilitate the movement of lithium ions back and forth between the cathode and the anode during charging and discharging cycles. Lithium ions are extracted from the cathode during charging, moving through the electrolyte to the anode. Conversely, during discharge, the ions return to the cathode.

The composition and structure of the cathode significantly influence the performance, energy density, and longevity of the battery pack. Materials with high energy density and stable cycling capabilities are desirable for cathode construction to maximize the overall efficiency and lifespan of the battery. Research and development efforts continually focus on improving cathode materials to enhance battery performance, increase energy storage capacity, and reduce costs in EV technology.

See also **Anode**.

CBET (Charge Before End Time)

CBET is a feature in EVs and associated charging infrastructure that enables users to schedule charging sessions to ensure that the vehicle reaches a specified SOC by a predetermined time, typically before a planned departure or subsequent usage. This feature allows EV owners to optimize charging times, take advantage of off-peak electricity rates, maximize battery health, and ensure that the vehicle is adequately charged when needed.

CCID (Charge Circuit Interrupting Device)

A CCID, also known as a charge circuit interrupter (CCI), is a safety device designed to protect against electrical hazards during EV charging. The CCID is integrated into EV charging equipment and automatically disconnects the charging circuit in case of a fault or hazardous condition. This could include a ground fault, leakage current, or insulation failure, which may pose a risk of electric shock or fire.

The CCID constantly monitors the flow of electric current during charging and detects any abnormalities or deviations from normal operating conditions. If a fault is detected, the CCID quickly interrupts the charging circuit, isolating the EV from the electrical supply and preventing potential safety hazards.

CCIDs are mandated by safety standards such as UL 2231-1 and UL 2202 for EV charging equipment, ensuring that EV charging infrastructure meets rigorous safety requirements. By incorporating CCIDs into charging stations and EVSE, manufacturers help ensure the safety of EV users, technicians, and bystanders during the charging process.

See also **Charge Interlocks**.

CCS (Combined Charging System)

The CCS is a standardized charging protocol for EVs that integrates AC and DC charging capabilities into a single connector and communication interface. Developed by major automakers and charging infrastructure providers, CCS aims to simplify EV charging and enable faster charging speeds across various EV models. CCS may employ a Type 1 (North American) or Type 2 (European) connector for AC charging. SAE J-1772 is a Type 1 format. CCS also supports Type 1 and 2 *Combo* connectors for DC fast charging.

This versatility enables CCS-equipped EVs to access various charging networks and infrastructure, including public charging stations, and private charging points.

CCS supports AC charging at power levels typically ranging from 3.7 to 22 kW, allowing EVs to recharge using standard household or commercial AC power sources. CCS enables high-power DC fast charging at power levels ranging from 50 to 350 kW or more, significantly reducing charging times and enabling long-distance travel with minimal downtime. Critical features of CCS include communication capabilities that allow EVs and charging stations to communicate and negotiate charging parameters, such as charging power, voltage, and current. This ensures efficient and safe charging while protecting the battery and optimizing charging speed.

CCS (Combined Charging System) Combo

CCS Combo is a standardized charging connector for EVs that integrate both AC and DC charging capabilities. The CCS Combo connector combines the Type 1 or 2 AC connector with additional DC pins, allowing EVs to charge using AC power from standard charging stations or DC power from high-power fast charging stations. Developed jointly by European and North American automakers, CCS Combo offers a single, compact port that accommodates multiple charging methods, making it a versatile solution for EV charging infrastructure [2, 3].

CD (Charge-Depleting) Mode

CD mode is an operational state in which an EV primarily relies on its battery pack to provide power for propulsion. During CD mode, the vehicle's electric motor draws energy from the battery to drive the wheels, allowing the EV to travel a certain distance or until the battery's SOC reaches a predetermined threshold [4].

In practical terms, the CD mode is typically engaged when the EV is in motion and requires power for acceleration, cruising, or climbing grades. During this mode, the battery's stored energy is gradually depleted as the vehicle consumes power to propel itself forward.

CD mode is commonly used in EVs with PHEV or BEV powertrains. In PHEVs, the vehicle may switch between CD and charge-sustaining modes, where a gasoline engine or other power source supplements the battery's energy to maintain propulsion. In BEVs, CD mode is the primary operating mode, as the vehicle relies solely on its battery for propulsion.

Efficiently managing battery energy use during CD mode allows EVs to optimize their driving range and overall energy efficiency, providing drivers with a reliable and sustainable mode of transportation. Additionally, advanced EMSs may employ strategies such as regenerative braking to recapture energy during deceleration and extend the vehicle's range further.

Cells

In EVs, cells refer to the individual units composing the battery pack. These cells are the fundamental building blocks of the battery system and are responsible for storing electrical energy in the form of chemical energy through electrochemical processes (Figure C.1).

Li-ion cells are the most common type of cell used in modern EV batteries due to their high energy density, lightweight, and relatively low self-discharge rate. Li-ion cells come in various chemistries, such as lithium cobalt oxide (LCO), lithium iron phosphate (LFP), lithium nickel manganese cobalt oxide (NMC), and lithium manganese oxide (LMO). An emerging technology replaces the liquid or gel electrolyte found in traditional Li-ion cells with a solid-state electrolyte. Solid-state cells offer potential advantages in terms of safety, energy density, and longevity but are still in the early stages of commercialization.

Like Li-ion cells, but with a polymer electrolyte instead of a liquid one, they can be shaped into flexible and lightweight forms, making them suitable for specific design applications. Although less common in modern EVs due to their lower energy density than Li-ion cells, nickel–metal hydride (NiMH) cells have been used in some HEVs and early EVs.

FIGURE C.1 A graphic of the bank of cells making up a battery pack.

IM Imagery/Shutterstock.com.

CEMS (Customer Energy Management Service)

CEMS refers to a suite of solutions and technologies offered by EV manufacturers or energy providers to optimize EV charging and usage while minimizing costs and maximizing energy efficiency for consumers. CEMS leverages smart charging, DR, energy storage, and advanced analytics to empower EV owners with tools for intelligent energy management [1].

CEMS integrates with EV charging infrastructure, smart meters, and energy management systems (EMSs) to monitor and control EV charging sessions. Some key functionalities of CEMS include smart charging, which dynamically adjusts charging rates based on grid conditions, energy prices, and user preferences to optimize charging times and minimize costs. Another function is in DR programs, which adjust charging schedules in response to grid signals or peak demand periods, contributing to grid stability and reducing electricity costs.

CEMS systems coordinate with home ESSs or V2G capabilities to store excess energy from renewable sources or discharge energy back to the grid during peak demand or emergencies. Finally, CEMS gives users data-driven insights into their energy consumption, charging patterns, cost savings, and environmental impact, enabling informed decision-making and behavior adjustments.

CHAdeMO Plug

The CHAdeMO plug is a connector used for charging EVs, primarily supporting DC fast charging. It was developed in Japan by the CHAdeMO Association, a consortium formed by major automotive manufacturers Toyota, Nissan, Mitsubishi, and Fuji Heavy Industries (SUBARU) together with Tokyo Electric Power Company (TEPCO) and has over 500 members from 50+ countries.

CHAdeMO is not only the name of the consortium but also the name of the DC fast charging system. "CHArge de MOve" is derived from the Japanese phrase "charge for moving." The pronunciation of the word "CHAdeMO" means a Japanese expression "How about having tea (while charging)?"

CHAdeMO plugs facilitate high-speed EV charging by delivering DC to the vehicle's battery, allowing faster charging times than standard AC charging methods. This type of plug supports up to 400 kW of power, enabling it to charge a typical EV battery from 0 to 80% in about 20 to 30 min, depending on the battery capacity, the battery status, and the charger's power output.

According to its latest specifications, CHAdeMO plugs are compatible with many EVs, especially those manufactured by Japanese automakers and other international brands that adopt this standard (Figure C.2). However, the popularity of the CHAdeMO plug varies by region. It is significantly used in Japan and other parts of Asia, but faces competition from other standards like CCS in Europe and North America, and newly emerging North American Charging System (NACS) in North America. CHAdeMO is unique in that it has been the only system that has interoperable bi-directional power transfer capabilities [V2H/vehicle to everything (V2X)] with products readily available.

FIGURE C.2 CHAdeMO chargers [5].

Courtesy of CHAdeMO.

(Continued)

FIGURE C.2 **(Continued)** CHAdeMO chargers [5].

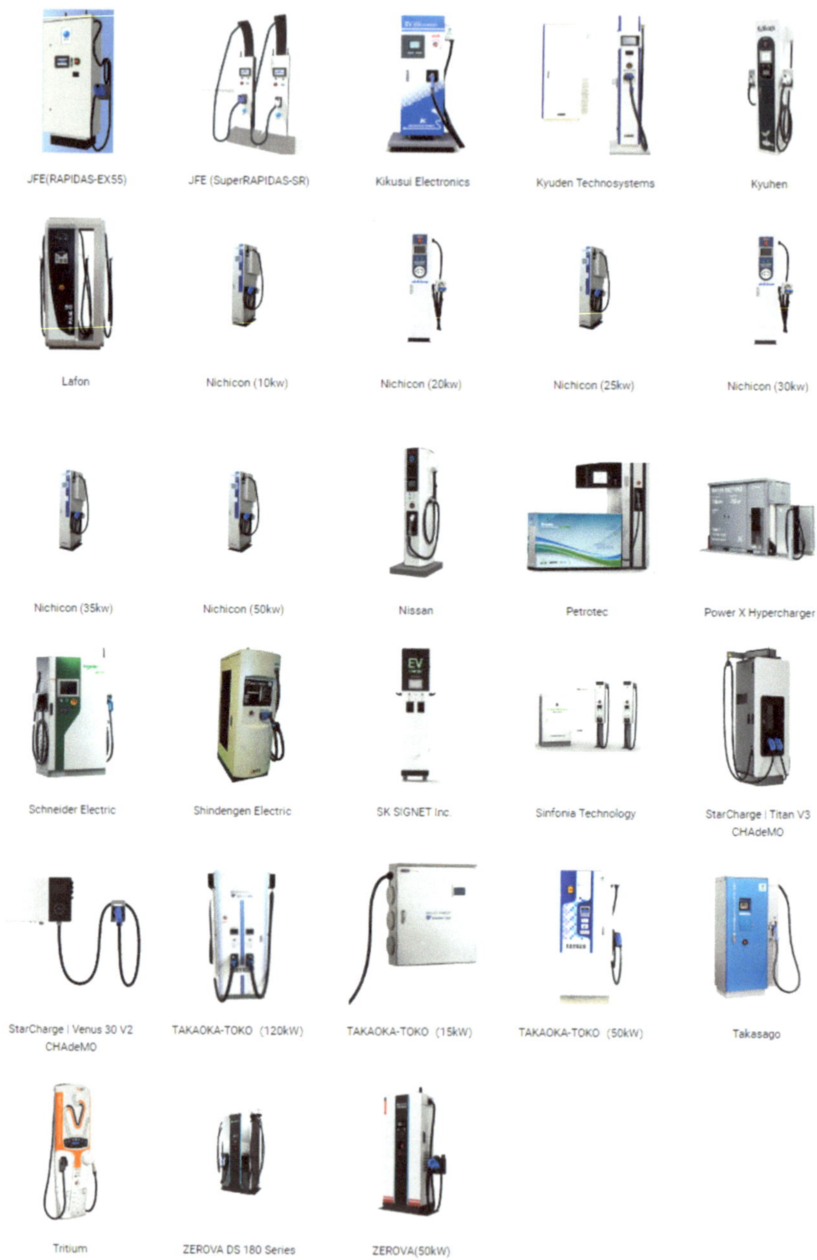

Courtesy of CHAdeMO.

Charge Connector

A charge connector, or a charging plug or charging inlet, is a vital component of EVs and PHEVs that facilitates the connection between the vehicle and a charging station or power source. It serves as the interface through which electrical power is transferred from the charging infrastructure to the vehicle's onboard charging system, allowing for the replenishment of the vehicle's battery.

Charge connectors come in various types and configurations, each designed to accommodate different charging standards and protocols. Common connector types include Type 1 (J1772), Type 2 (Mennekes), CHAdeMO, CCS, and Tesla's proprietary connector [2].

Charge connector compatibility varies depending on the EV or PHEV model and the type of charging structure available. Some vehicles may have multiple charging ports or adapters to accommodate different connector types. Charge connectors often incorporate safety features such as locking mechanisms, grounding pins, and communication protocols to ensure safe and secure charging operations. These features help prevent electric shock, overheating, and other hazards during the charging process.

Charge Contactors

Charge contactors are electrical components used in EVs to control the flow of electrical current between the vehicle's battery pack and the charging infrastructure. These contactors act as switches that open or close the circuit between the battery and the charging port, allowing or interrupting the flow of electricity during the charging process. Charge contactors ensure safe and efficient charging operations by providing isolation and protection against electrical faults, such as overcurrent or short circuits. They are typically controlled by the vehicle's charging system and are activated when initiating or terminating the charging process. Charge contactors are designed to handle high currents and voltages encountered during EV charging, and they are essential components of the charging system's safety and control mechanisms [2].

Charge Controller

A charge controller, also known as a charging controller or charge management system, is a vital component of EV charging infrastructure responsible for regulating the flow of electric current during the charging process. It manages the charging process efficiently, optimizes battery performance, and ensures the charging system's safety and reliability.

The charge controller monitors and controls the voltage levels during the charging process to prevent overcharging, which can damage the battery and reduce its lifespan. It regulates the electric current supplied to the battery, ensuring that it remains within limits and preventing excessive heating or stress on the cells. Many charge controllers incorporate temperature sensors to monitor the temperature of the battery and charging equipment. This allows the controller to adjust the charging parameters to prevent overheating and ensure optimal charging efficiency.

Charge controllers may also provide information about the battery's SOC, allowing users to monitor the charging process' progress and estimate the remaining charging time. Charge controllers often include built-in safety features such as short-circuit protection, overcurrent protection, and overvoltage protection to prevent battery and charging equipment damage in case of a fault or malfunction.

Charge Coupler Optional Markings

Charge coupler optional markings refer to additional visual or textual identifiers on or near the EV and HV charging coupler. These markings provide critical information to users and technicians, enhancing safety, compatibility, and ease of use during charging [6].

1. **Compatibility Information:** It indicates the types of charging stations compatible with the vehicle's charging system (e.g., Type 1, Type 2, CHAdeMO, and CCS).
2. **Safety Instructions:** They involve providing specific safety-related information to prevent misuse or hazards during the charging process, such as voltage, current capacity, and environmental conditions under which the coupler can safely operate.
3. **Technical Specifications:** They involve detailed technical data relevant to the vehicle's charging capabilities, such as maximum power input and communication protocol compatibility.

These markings are typically found directly on the charge coupler, its housing, or the vehicle's charging compartment, ensuring users' visibility and accessibility.

In many regions, standards and regulations may dictate the content, format, and visibility of charge coupler optional markings to ensure uniformity and compliance across different manufacturers and models.

These markings help users identify the correct type of charging station, reducing the risk of damage to the vehicle or the charging infrastructure. They also provide essential safety tips that protect users from potential electric shock or other electrical hazards [2].

Charge Flap

The charge flap, also known as the charging port or charging inlet, is a component of an EV that provides access to the vehicle's battery for charging purposes (Figure C.3). Typically located on the vehicle's exterior, the charge flap serves as the interface through which charging cables or connectors are connected to the EV for recharging its battery. Depending on the vehicle's make and model, it is designed to accommodate various charging connectors, such as Type 1, Type 2, CCS, CHAdeMO, or Tesla's proprietary connector [7]. The charging flap often includes locking mechanisms, indicator lights, and weatherproof seals to ensure secure and reliable connections and protect the charging port from environmental elements.

FIGURE C.3 An example of a charge flap for an EV.

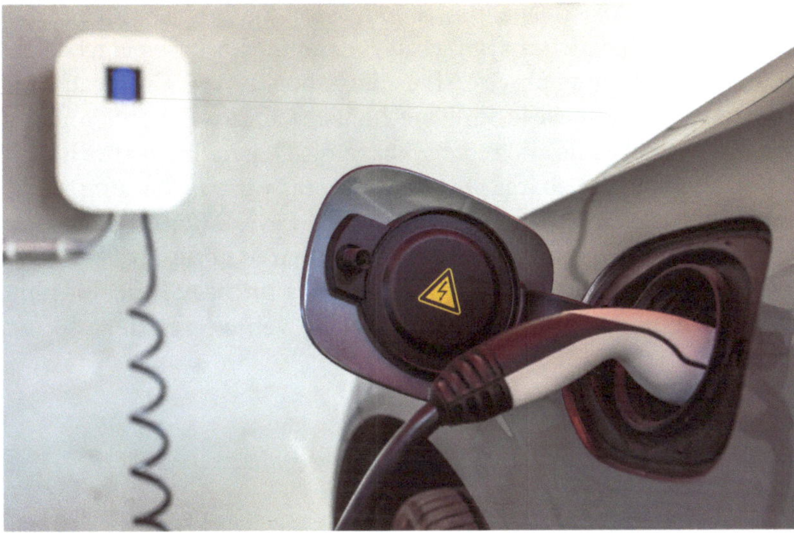

Canetti/Shutterstock.com.

Charge Interlocks

Charge interlocks are safety features implemented in EVs and EVSE, also known as charging stations, to prevent hazardous conditions and ensure safe charging operations [2]. These interlocks typically consist of electronic or mechanical mechanisms that control the flow of electricity between the EV and the charging station and ensure proper connection and disconnection of the charging cable.

The primary function of charge interlocks is to prevent unintended energizing of the charging cable and connectors, which could pose risks of electrical shock or damage to equipment. They are designed to activate and deactivate automatically during the charging process, based on various conditions such as proper alignment of connectors, engagement of safety mechanisms, and communication between the EV and the charging station.

Charge interlocks may include features such as plug detection, which ensures that the charging plug is inserted correctly and securely into the EV's charging port before electricity flow is initiated. Interlocks also detect the presence of objects or obstacles near the charging connector to prevent accidental contact or interference during charging.

Charge interlocks secure the charging cable and connectors to prevent accidental disconnection or tampering during charging. The interlocks include communication protocols, establishing communication between the EV and the charging station to verify compatibility, authenticate users, and exchange status information. Lastly, safety overrides allow for emergency shutdown or interruption of charging operations in case of malfunctions, safety hazards, or abnormal conditions.

Charge Port

The charge port of an EV is a designated access point where the vehicle's battery can be connected to an external power source for charging (Figure C.4). It is the interface through which charging cables or connectors are plugged into the EV to initiate the charging process. The charge port is typically located on the vehicle's exterior. It may have locking mechanisms, indicator lights, and weatherproof seals to ensure safe and reliable charging. Depending on the EV model and charging capabilities, charge ports may support various charging standards and connector types, including Level 1 (120 V), Level 2 (240 V), and DC fast charging [7]. The design and placement of the charge port may vary between different EV models and manufacturers. Still, its primary function remains consistent: to facilitate the recharging of the vehicle's battery to enable continued operation.

FIGURE C.4 An example of the vehicle charge port.

kckate16/Shutterstock.com.

Charge Start Time

The charge start time refers to the specific time at which the charging process of an EV begins. EV owners can schedule the start time for charging using various methods, such as through the vehicle's onboard charging timer or via a smartphone app connected to the charging station. Scheduling the charge start time allows EV owners to take advantage of off-peak electricity rates, optimize charging times to coincide with periods of lower grid demand, or ensure that the vehicle is fully charged and ready for use at a specific time. By setting the charge start time, EV owners can manage their charging sessions efficiently, reduce electricity costs, and maximize the convenience of EV ownership.

Charge Status Indicator

A charge status indicator is a visual or digital display on an EV dashboard or charging interface that provides real-time information about the vehicle's battery's current charge (SOC) state. The indicator typically shows the remaining battery capacity as a percentage or a graphical representation, allowing the driver to monitor the battery's charge level at a glance [8]. Charge status indicators may include additional information such as estimated driving range based on the current charge level, charging status during charging sessions, and alerts for low battery levels or charging errors. These indicators are essential for EV drivers to plan their trips effectively, manage charging stops, and ensure sufficient battery capacity for their intended journeys.

Charge Stock Time

Charge stock time is when an EV remains connected to a charging station without actively charging. This period typically occurs when an EV is plugged into a charging station but is not actively receiving electrical power to recharge its battery. Charge stock time may occur for various reasons, including when a charging station is occupied or when an EV driver plugs in their vehicle but delays starting the charging session, or when an EV remains connected to a charging station after completing its charging cycle. Minimizing charge stock time is essential to optimize the availability and utilization of charging infrastructure, allowing more EVs to access charging stations efficiently and reducing waiting times for charging [2].

Charge-Sustaining Mode

Charge-sustaining mode refers to the operational state of a PHEV or extended range EV (EREV) in which the vehicle's ICE is engaged to maintain or sustain the SOC of the vehicle's battery pack. In this mode, the ICE generates power to propel the vehicle while also providing electrical energy to recharge the battery pack, thereby preventing the battery from discharging below a predetermined threshold. Charge-sustaining mode allows PHEVs and EVs to operate for extended distances beyond the AER by utilizing electric and gasoline power sources, providing increased flexibility and range for drivers. This mode is commonly used during highway driving or under heavy load conditions when the vehicle's power demand exceeds the battery's capacity alone.

For more information, see **J1715** [3].

Charge with Inverter

Charge with inverter refers to a feature in certain EVs that enables the vehicle's battery pack to be charged using an external power source, typically an AC power outlet [9]. The inverter also converts the AC input to DC, suitable for battery charging. This feature is handy when charging an EV from a standard household electrical outlet, which provides AC power [3]. The onboard inverter within the EV converts the incoming AC power to DC power, which is then used to charge the vehicle's battery pack. Charge with inverter functionality allows EV owners to conveniently recharge their vehicles using various power sources, including residential outlets, public charging stations, and portable generators, without additional external equipment or adapters. It enhances the flexibility and accessibility of EV charging options, enabling drivers to recharge their vehicles wherever AC power is available.

Charger

A functional component that can be located on or off the vehicle that supplies electrical power at the correct voltage and current levels for battery charging.

For more information, see **J1715** [3].

Charger Anxiety

Charger anxiety, sometimes referred to as range anxiety, is a term used to describe the concern or stress experienced by EV drivers about the availability of charging infrastructure and the potential for running out of battery charge during a journey. EV drivers may experience charger anxiety when they perceive a lack of accessible charging stations along their planned route or when they are uncertain about the battery range of their vehicle compared to the distance they need to travel. This anxiety can impact the confidence and convenience of EV ownership and may influence driving behaviors, route planning, and overall adoption of EVs. Charger anxiety highlights the importance of expanding and improving charging infrastructure to alleviate concerns and promote the widespread adoption of EVs.

Charging

Charging refers to replenishing the electrical energy stored in an EV or PHEV battery (Figure C.5). During charging, electric power is transferred from an external power source, such as a charging station or electrical outlet, to the vehicle's battery pack, which is stored for later use in powering the vehicle's electric motor(s). Charging can occur through various methods and charging levels, including Level 1 (120V AC), Level 2 (240V AC), and DC fast charging (DC), each offering different charging speeds and compatibility with EVs [2, 7].

FIGURE C.5 Typical closed-loop charging system [7].

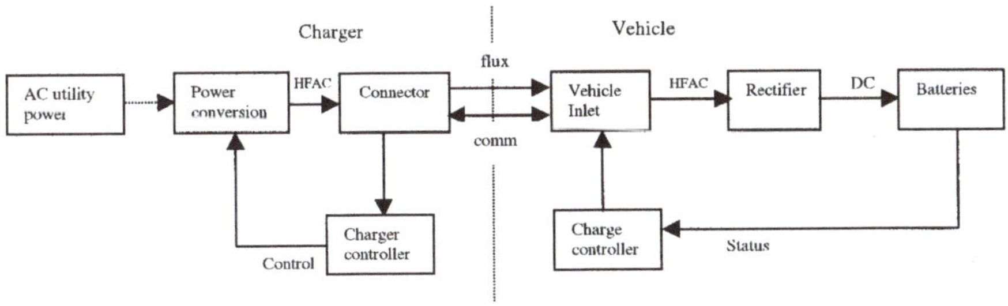

Charging Capacity

Charging capacity refers to the maximum rate at which an EV battery can accept electrical energy during charging. It is typically measured in kilowatts (kW) and represents the speed at which the battery can be replenished with energy from an external power source, such as a charging station. A higher charging capacity allows for faster charging times, reducing the time required to recharge the EV battery and increasing the convenience and flexibility of EV ownership. Charging capacity is influenced by factors such as the battery's design, chemistry, temperature, and SOC, as well as the capabilities of the charging infrastructure and charging equipment. Understanding the charging capacity of an EV battery is essential for selecting appropriate charging solutions and optimizing the charging experience for EV owners.

Charging Complete Notification

Charging complete notification refers to a signal or alert provided by an EV charging system to indicate that the charging process has finished and the EV battery is fully charged. This notification may be delivered through various means, including visual indicators on the charging station, notifications on the EV charging app or dashboard display, audible alerts, or notifications sent to the EV owner's smartphone or other connected devices. The purpose of the charging complete notification is to inform the EV owner that their vehicle is fully charged and ready for use, allowing them to disconnect the charging cable and continue their journey. This notification helps ensure efficient use of charging infrastructure and prevents overcharging of the EV battery, optimizing battery health and prolonging its lifespan.

Charging Infrastructures

Charging infrastructure refers to the network of facilities, equipment, and technologies designed to support charging for EVs. This infrastructure includes various charging stations, also known as EVSE, which provide electric power for recharging EV batteries. Charging infrastructure encompasses multiple charging solutions, including Level 1, Level 2, and DC fast charging stations, each offering different charging speeds and capabilities. Additionally, charging infrastructure may include components such as charging connectors, cables, billing and payment systems, and software platforms for managing and monitoring charging operations. The development and expansion of charging infrastructure are essential for facilitating the widespread adoption of EVs by providing convenient, accessible, and reliable charging solutions for EV owners at home and in public locations.

For more information, see "A Cloud-Based Simulation and Testing Framework for Large-Scale EV Charging Energy Management and Charging Control" [1].

Charging Input

Charging input refers to the method or interface through which EVs receive electrical energy from an external power source to recharge their battery packs. It encompasses the charging connectors, ports, and protocols for connecting the EV and the charging infrastructure. Charging inputs are designed to accommodate different charging levels and speeds, ranging from standard household outlets to specialized charging stations with higher-power output capabilities.

In EV terminology, charging input often refers to the physical connector or port on the vehicle that allows it to be connected to an external charging source. This connector may vary depending on the charging standard or EV-supported protocol [2].

Charging Network

A charging network refers to a system of interconnected charging stations or infrastructure to supply electric power for recharging EVs. These charging stations are strategically located in various public and private locations, including highways, parking facilities, shopping centers, and residential areas, to provide convenient and accessible charging options for EV owners. The purpose of a charging network is to support the widespread adoption of EVs by supplying EV owners with reliable and efficient charging solutions, thereby reducing range anxiety and promoting the transition to clean and sustainable transportation. Charging networks may be operated by government agencies, utilities, automotive manufacturers, or third-party service providers and often incorporate features such as online maps, mobile apps, and payment systems to facilitate ease of use and accessibility for EV drivers.

Charging Output

Charging output refers to the rate at which electrical energy is delivered to an EV battery during the charging process. It is typically measured in kilowatts (kW) and represents the power level at which the EV charger supplies electricity to the vehicle's battery pack. A higher charging output allows faster charging times, enabling the EV to replenish its battery capacity more quickly. Charging output depends on various factors, including the type of charging station used (e.g., Level 1, Level 2, DC fast charging), the charging infrastructure's capabilities, and the EV's onboard charging system. Understanding the charging output is essential for EV owners to estimate charging times accurately and plan their charging sessions effectively based on their driving needs and schedule.

Charging Point

A charging point, also known as a charging station or EV charging station, is a designated location equipped with infrastructure for supplying electric power to recharge the battery of an EV. Charging points come in various forms, including wall-mounted units, standalone pedestals, or integrated into existing infrastructure such as parking meters or light poles. These points are equipped with connectors and charging equipment compatible with different types of EVs and charging standards. Charging points may provide various levels of charging, including Level 1 (standard household outlet), Level 2 (AC charging), and Level 3 (DC fast charging) [2]. Each offers varying charging speeds and capabilities. Charging points are typically installed in public locations such as parking lots, shopping centers, rest areas, and highways to provide convenient and accessible charging solutions for EV owners. Additionally, charging points may be installed in private settings such as residential garages, workplaces, and fleet depots to support home charging and commercial charging.

Charging Protocol Level Detection by Supply Equipment

Charging protocol level detection by supply equipment refers to the capability of charging infrastructure, such as EVSE or charging stations, to detect the communication protocol level supported by the EV connected for charging. This feature allows the charging equipment to communicate effectively with the EV and adjust the charging parameters accordingly, ensuring safe and efficient charging sessions. Charging protocol level detection by the charging station relieves the worry about selecting the correct charging protocol. It is incredibly convenient and ensures that the EV charges efficiently every time.

Charging Protocol Level Signaling

Charging protocol level signaling, is a communication protocol used in EV charging systems to facilitate communication between the EV and the charging station. This signaling protocol allows the EV and the charging station to exchange information regarding charging capabilities, power requirements, and other relevant parameters to ensure safe and efficient charging.

During the charging process, the EV and the charging station engage in charging protocol level signaling to establish communication and negotiate the optimal charging parameters based on the capabilities of both the vehicle and the charging infrastructure. This ensures safe, efficient, and reliable charging for EVs.

Charging Rate

The EV charge rate refers to the speed at which an EV can recharge its battery pack. It is typically measured in kilowatts (kW) or miles of range added per hour of charging (mi/h). A higher charge rate means the EV can replenish its battery capacity more quickly, reducing the time required for charging. The charge rate can vary depending on the EV's battery capacity, charging infrastructure, and the type of charging connector used [2]. Understanding the charge rate is essential for EV owners to plan their charging sessions effectively and optimize their driving experience.

Charging Stations

An EV charge station, also known as an EV charging station or EVSE, is a specialized infrastructure facility designed to supply electric power for recharging EVs (Figure C.6). Charge stations come in various types and configurations, each providing different charging speeds and compatibility with EVs. These stations are equipped with charging connectors and control systems that allow EV owners to connect their vehicles for charging. Charge stations may be installed in public locations such as parking lots, rest areas, commercial establishments, and private settings such as residential garages and workplaces. They play a crucial role in supporting the widespread adoption of EVs by providing convenient and accessible charging solutions for EV owners.

For more information, see "A Cloud-Based Simulation and Testing Framework for Large-Scale EV Charging Energy Management and Charging Control" [1].

FIGURE C.6 Charging stations are like gas stations for EVs.

Charging Time

This equation assumes that the charging process is linear and does not account for any tapering of charge rate that might occur as the battery approaches full charge, especially in the case of fast charging. For more precise calculations, factors like the efficiency of the charging system, ambient temperature, and specific battery technology should also be considered.

$$T = \left(\frac{C \times \left(SOC_{end} - SOC_{start} \right)}{P} \right) \times \eta$$

where

T is the charging time
C is the battery capacity in kilowatt-hours (kWh)
SOC_{end} is the desired SOC at the end of charging (expressed as a decimal)
SOC_{start} is the initial SOC before charging begins (expressed as a decimal)
P is the power of the charger in kilowatts (kW)
η is the charging efficiency (less than 1, accounting for energy lost during the charging process)

Chassis Ground

Chassis ground, also known as earth ground or vehicle ground, refers to the electrical connection between an EV's chassis and Earth's surface (Figure C.7). In EVs, the chassis serves as the standard reference point for electrical circuits and components, providing a stable and low-impedance path to dissipate electrical charges and ensure safety. Chassis ground is essential for establishing a stable electrical reference point, preventing electrical shocks, and protecting against electrical faults or malfunctions. It helps to ensure the safe operation of electrical systems and components in the vehicle and provides a path for dissipating static electricity and minimizing EMI [10]. Chassis ground is established by connecting various components, such as the battery, powertrain, and electrical systems, to the vehicle's metal frame or structure [11].

FIGURE C.7 Typical xEV high-voltage system [11].

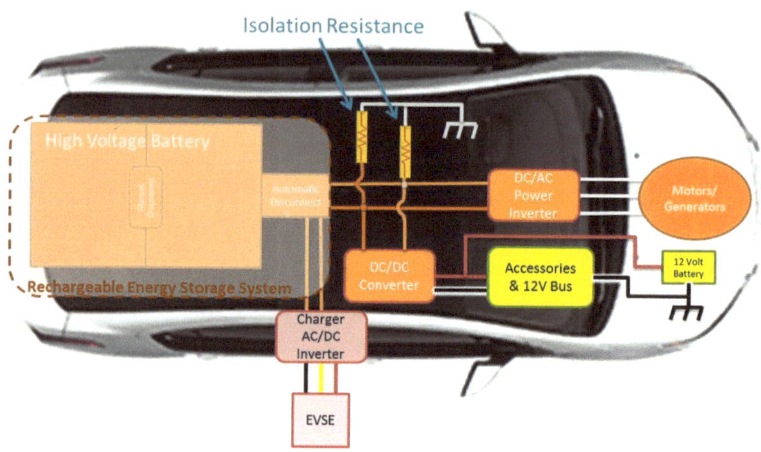

Circuit Interrupters

Circuit interrupters (automatic disconnect), also known as circuit breakers or interrupter switches (Figure C.8), are safety devices designed to protect electrical circuits and equipment from damage caused by overload, short circuits, or ground faults [11]. Circuit interrupters play a critical role in ensuring the safety and reliability of EV electrical systems [2].

Circuit interrupters are commonly used in various EV systems, including the battery, charging system, and onboard electronics. They provide an additional layer of safety and protection, complementing other safety features such as fuses, ground-fault protection devices, and insulation monitoring systems.

See also **Charge Interlocks**.

FIGURE C.8 A proximity detection circuit is an example of a circuit interrupter [2].

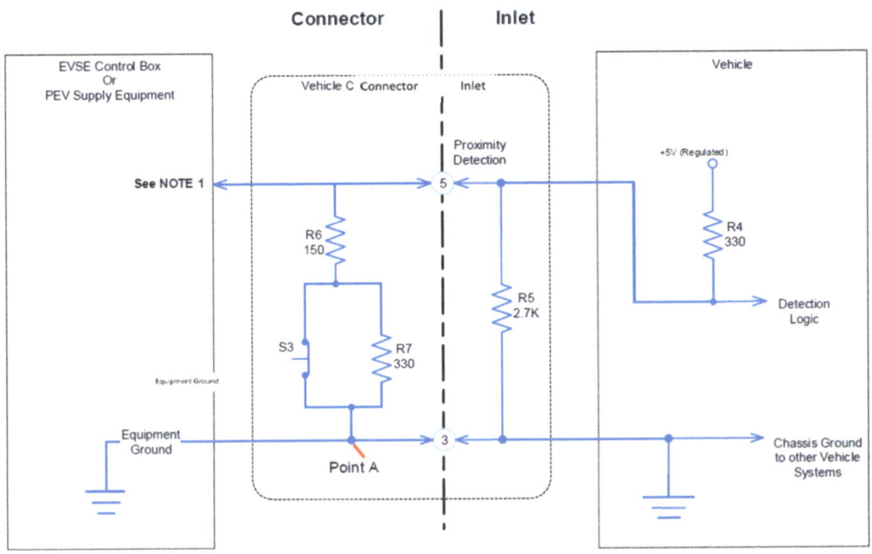

Class 9 Hazardous Material

Class 9 hazardous material refers to substances that present a potential risk to health, safety, property, or the environment during transportation, handling, or storage (Figure C.9). In the context of EVs, Class 9 hazardous materials may include Li-ion batteries and other components containing hazardous materials, such as electrolytes or flammable materials [12]. These materials are classified as hazardous due to their potential to cause fires, explosions, or other hazards if not properly handled or managed. Transporting, storing, or disposing of Class 9 hazardous materials requires compliance with regulations and safety standards established by regulatory authorities, such as the Department of Transportation (DOT) in the United States (US), or the International Maritime Organization (IMO) internationally. Proper handling, packaging, labeling, and transportation procedures are essential to minimize risks and ensure the safe and environmentally responsible management of Class 9 hazardous materials associated with EVs.

FIGURE C.9 An example of marking when shipping Class 9 materials [12].

Courtesy of DOT.

Climate Control

Climate control systems in EVs and HVs refer to the integrated mechanisms that maintain the desired temperature and air quality within the vehicle's cabin. This system encompasses HVAC functions.

Climate control systems in EVs and hybrids differ from those in traditional ICE vehicles primarily due to their energy source. In EVs and hybrids, the system is powered by the vehicle's main battery rather than an engine-driven compressor. This configuration can impact the vehicle's overall range and efficiency, as climate control functions are significant electrical power consumers.

Code Set Connector

A code set connector, also known as a diagnostic connector or onboard diagnostics (OBD) connector, is a standardized interface used in EVs and other vehicles to facilitate communication between the vehicle's electronic control systems and external diagnostic equipment (Figure C.10) [15]. This connector enables technicians and service personnel to access information about the vehicle's operation, performance, and diagnostic trouble codes (DTCs) for troubleshooting and maintenance [16].

In most modern vehicles, including EVs, the code set connector follows a standardized protocol like the OBD-II standard [17]. It provides access to a standardized set of data parameters and diagnostic codes, allowing compatible diagnostic tools to communicate with the vehicle's onboard computer systems. By connecting diagnostic equipment to the code set connector, technicians can retrieve real-time data, monitor system status, and diagnose issues related to the vehicle's electrical systems, including the battery, motor, charging system, and other electronic components.

Using a code set connector in EVs enables efficient diagnosis and maintenance of electric propulsion systems, ensuring optimal performance, reliability, and safety. Additionally, the standardized interface promotes interoperability among diagnostic tools and facilitates compatibility with a wide range of vehicles, simplifying servicing and repair procedures for EV owners and service technicians.

FIGURE C.10 Connector orientation (mounting limits) [15].

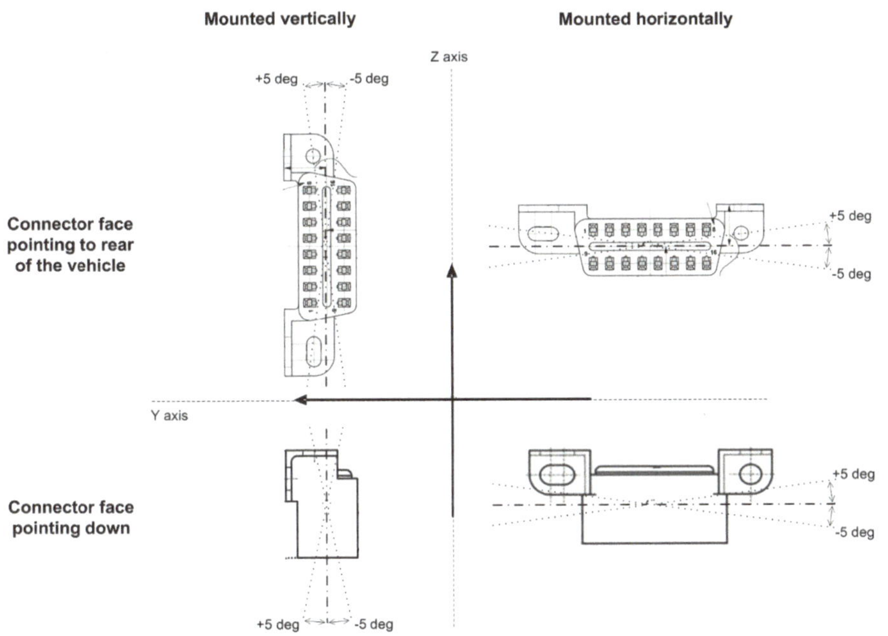

Courtesy of NIST.

Commercial EV

A commercial EV is designed and used primarily for commercial or business purposes, such as transportation of goods, delivery services, passenger transport, or specialized applications in construction, agriculture, and logistics industries. Commercial EVs encompass various vehicle types, including vans, trucks, buses, utility vehicles, and specialty vehicles, all powered by electric propulsion systems.

These vehicles typically feature battery electric or PHEVs drivetrains, which enable them to operate solely on electric power or in combination with an ICE or range extender. Commercial EVs offer several advantages over traditional ICE vehicles, including reduced operating costs, lower emissions, quieter operation, and potential eligibility for incentives and subsidies to promote clean transportation.

Commercial EV adoption is gaining traction globally as businesses seek to reduce their carbon footprint, comply with environmental regulations, and lower their total cost of ownership. As technology continues to advance and charging infrastructure expands, the market for commercial EVs is expected to grow, offering a sustainable and efficient alternative to conventional diesel and gasoline-powered vehicles in various commercial applications.

Conductive Charge Coupler

A conductive charge coupler, or a charging connector or plug, is used in EV charging systems to establish a conductive electrical connection between the EV and the charging station. The charge coupler typically consists of a male connector on the vehicle side and a corresponding female receptacle on the charging station side. It allows for transferring electrical power from the charging infrastructure to the EV's battery for charging purposes.

Conductive charge couplers come in various shapes, sizes, and configurations, depending on factors such as the charging protocol (e.g., AC charging DC fast charging), the power rating of the charging station, and regional standards and regulations.

The conductive charge coupler facilitates EVs' safe and efficient charging, providing a reliable and standardized interface for connecting EVs to charging infrastructure. As the adoption of EVs continues to grow, interoperability and compatibility of charge couplers across different EV models and charging stations are essential to support widespread EV charging infrastructure deployment and enable seamless charging experiences for EV drivers.

Conductive Coolant

Conductive coolant, or electrically conductive coolant, is a specialized fluid used in EVs to regulate the temperature of the vehicle's battery pack and other electrical components. Unlike traditional engine coolants used in ICE vehicles, conductive coolants are formulated to possess electrical conductivity properties. This allows it to effectively dissipate heat generated during battery operation and electrical system use, helping maintain optimal operating temperatures and prolonging the lifespan and performance of the EV's battery and electrical components. Conductive coolant circulates through a cooling system comprising coolant loops, pumps, radiators, and heat exchangers to transfer heat away from critical components and dissipate it into the surrounding environment. Conductive coolant is crucial in ensuring EVs' safe and reliable operation by efficiently managing thermal loads and preventing overheating.

See also **Thermal Management System**.

Connection Cycle

A connection cycle in the context of EVs refers to plugging and unplugging the EV from a charging station or power source. A connection cycle occurs each time an EV is connected to or disconnected from a charging station. Connection cycles are essential for the durability and reliability of the charging infrastructure and the EV's charging port.

Understanding connection cycles is crucial for assessing the wear and tear on charging equipment, such as connectors and cables. Frequent cycles can lead to mechanical wear, corrosion, and electrical degradation over time. Manufacturers of charging stations and EVs often specify the expected number of connection cycles their products can withstand before requiring maintenance or replacement.

Monitoring connection cycles and maintaining charging equipment properly can help ensure the safety, efficiency, and longevity of EV charging infrastructure, minimizing downtime and maximizing the lifespan of components. Additionally, advancements in connector design and materials aim to increase durability and withstand a more significant number of connection cycles, improving the reliability of EV charging systems.

Connector

A connector is a component which terminates conductors for the purpose of providing connection and disconnection to a suitable mating component. For the purposes of charging an EV, this component creates a coupling from the EVSE to the vehicle, to transfer energy and to exchange information.

For more information, see **J1715** [3].

Contactless Power Transfer

See **Inductive Charging System**.

Contactor

A contactor is an electrical switch used in EVs and HVs to control the flow of high-voltage power between components, such as the battery pack, motor controllers, inverters, and other electrical systems (Table C.1). It is a critical component that facilitates the smooth and safe operation of the vehicle's powertrain by managing the switching and routing of electrical currents.

The primary function of a contactor is to establish or interrupt electrical connections within the vehicle's power distribution system. When activated, the contactor closes its contacts, allowing high-voltage current to flow from the battery pack to the motor controllers, enabling propulsion. Conversely, when deactivated, the contactor opens its contacts, interrupting power flow and effectively disconnecting the high-voltage circuit, ensuring safety during maintenance, servicing, or in the event of a fault.

There are several types of contactors, starting with the main contactor. This contactor controls the primary power circuit between the battery pack and the motor controllers, which are responsible for powering the electric motor during vehicle operation. Next is the pre-charge contactor, which gradually applies voltage to the system before engaging the main contactor, reducing inrush currents and minimizing stress on components. The service disconnect contactor provides a manual or automatic means of disconnecting high-voltage power for maintenance or emergencies, enhancing safety for technicians and first responders. Lastly, the auxiliary contactor manages the vehicle's secondary electrical circuits or subsystems, such as cooling systems, auxiliary power outlets, or regenerative braking systems.

TABLE C.1 Contact size electrical rating of charge coupler [2].

Contact #	Function	Size (mm)	Current rating (Amps)	Voltage rating
1	Power	3.6 diameter	Up to 80 A	300 min, up to 600 VAC/VDC
2	Power	3.6 diameter	Up to 80 A	300 min, up to 600 VAC/VDC
3	Equipment/ chassis ground	2.8 diameter	Per coupler rating as defined in Ground Path Current Test of UL2251	
4	Control pilot	1.5 diameter	2 A	30 VDC
5	Proximity	1.5 diameter	2 A	30 VDC

Courtesy of EPA.

Contracted Curtailment Amount

The contracted curtailment amount refers to the predetermined quantity of energy in an EV charging station or a fleet operator's agreement to curtail or reduce under certain circumstances. This term is particularly relevant in the co-response programs or agreements between EV charging infrastructure providers and grid operators.

As the adoption of EVs continues to grow, managing their charging patterns becomes crucial for maintaining grid stability and reliability. Contracted curtailment amount signifies the agreed-upon energy consumption limit that an EV charging station or a fleet is willing to curtail upon request by the grid operator or energy aggregator.

For instance, in a DR scenario where grid congestion or peak demand is anticipated, grid operators may call upon charging stations to reduce their electricity consumption temporarily. The contracted curtailment amount specifies the maximum extent to which charging operations can be scaled back without compromising the needs of EV owners or fleet operators.

Control Pilot

The control pilot is crucial in EV charging systems (**Figure C.11**). It serves as a communication interface between the charging station and the vehicle. It facilitates the negotiation of charging parameters such as voltage, current, and charging rate, ensuring safe and efficient charging sessions [6].

FIGURE C.11 Example of control pilot circuit and interfaces [2].

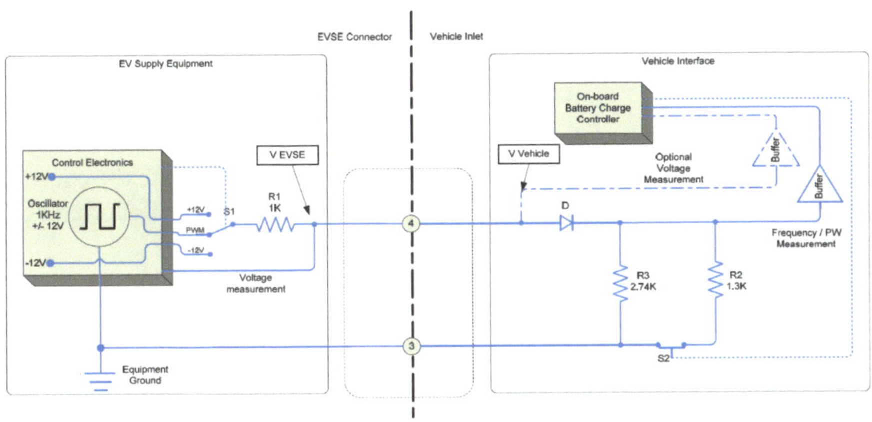

Courtesy of DOE.

The control pilot sends signals between the EV and the charging station, enabling the two devices to connect and exchange necessary information. This communication process allows the charging station to identify the vehicle, verify compatibility, and initiate charging at optimal settings.

The control pilot enables the following key functionalities:

- **Safety Checks:** The control pilot conducts safety checks to ensure that the EV and the charging station are compatible, that the electrical connections are secure, and that conditions for safe charging are met, such as proper grounding and the absence of faults.
- **Protocol Negotiation:** Once safety checks are successful, the control pilot facilitates protocol negotiation between the EV and the charging station. This includes determining the charging mode (e.g., AC or DC), maximum charging power, and communication protocols (e.g., **ISO 15118** for intelligent charging).
- **Power Delivery Coordination:** After protocol negotiation, the control pilot coordinates the power delivery process, instructing the charging station to supply the appropriate voltage and current based on the EV's charging capabilities and selected charging mode.

Control pilots can vary based on the charging protocol and standards of different regions and manufacturers. Control pilots support different charging modes, such as AC and DC, depending on the charging infrastructure and vehicle capabilities. They ensure that the charging process aligns with the requirements of the EV's battery system for safe and efficient operation.

Control pilots incorporate safety mechanisms to prevent overcharging, overheating, and other potential hazards during the charging process. These features include protocols for continuously monitoring charging parameters, emergency shutdown procedures, and communication of fault conditions between the vehicle and the charging station.

For more information, see **J1715** [3].

Control Pilot Tolerance

Control pilot tolerance refers to the permissible range within which the communication signal between an EV and a charging station's control pilot circuit can vary while maintaining a reliable connection.

Communication between the vehicle and the charging station is essential for safe and efficient power transfer during the charging process of EVs. The control pilot circuit is a crucial component that facilitates this communication by exchanging signals to initiate and control the charging process.

Control pilot tolerance specifies the acceptable deviation from the standard communication signal that the EV and charging station can tolerate without interrupting or compromising the charging process. This tolerance ensures compatibility between EVs and charging stations, allowing for reliable charging across various manufacturers and models.

A narrow control pilot tolerance ensures precise communication between the EV and the charging station, minimizing the risk of errors or malfunctions during the charging process. Conversely, a broader tolerance accommodates slight variations in the communication signal, which can benefit interoperability between different generations of EVs and charging infrastructure.

Standardization organizations, such as the International Electrotechnical Commission (IEC) and SAE International, define control pilot tolerance levels to ensure consistency and compatibility in EV charging protocols. These standards help manufacturers design EVs and charging stations that adhere to specific communication requirements, promoting seamless integration and interoperability within the EV ecosystem.

Control Sequence

In EVs and HVs, a control sequence refers to the specific set of programmed instructions or algorithms used to regulate the operation of the propulsion system (Figure C.12). This includes integrating and managing components such as the electric motor, ICE (in hybrids), battery, and other ancillary systems.

FIGURE C.12 An example overview of motor control [18].

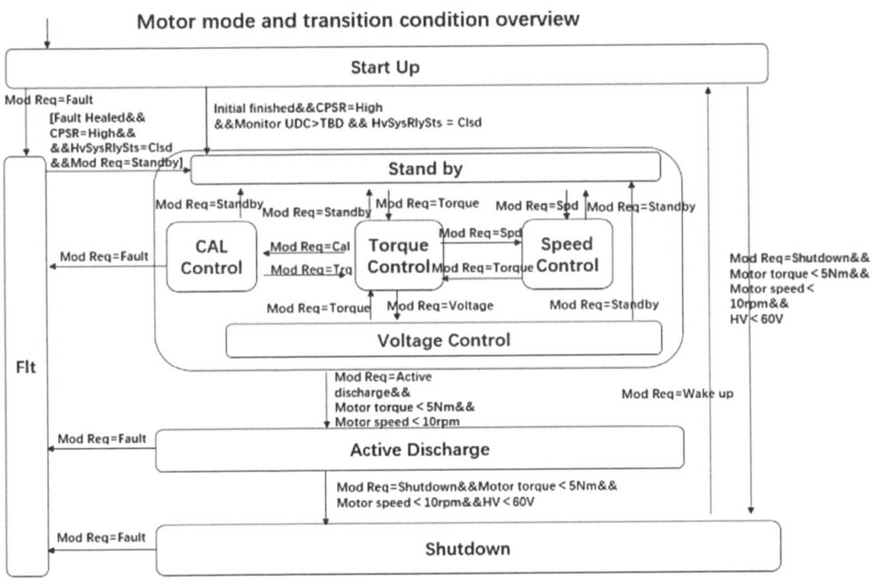

The control sequence ensures optimal vehicle performance, efficiency, and safety by managing how and when different power sources are utilized. For example, the control sequence in HVs determines when to switch between the electric motor and the ICE or when to use both for maximum efficiency.

Components Managed:

1. **Battery Management:** Monitors and controls the battery's SOC, health, and temperature to optimize longevity and performance.
2. **Torque Distribution:** Regulates the torque distribution between the electric motor and, in hybrids, the ICE to ensure smooth operation and transition.
3. **Regenerative Braking:** This system manages the capture of kinetic energy during braking, converting it into electrical energy that is stored in the battery.

4. **Thermal Management:** Oversees the cooling and heating systems to maintain optimal operating temperatures for the engine, motor, and battery.

The control sequence balances fuel efficiency, emissions reduction, and driving performance. Advanced control sequences use real-time data and predictive algorithms to adapt to changing driving conditions and preferences.

Cooperative Regenerative Braking

Cooperative regenerative braking refers to a sophisticated braking system employed in EVs that harnesses the kinetic energy during deceleration or braking and converts it into electrical energy (Figure C.13) [19]. Unlike traditional regenerative braking systems, which primarily rely on the vehicle's motion to generate power, cooperative regenerative braking extends its functionality to interact with other vehicles or infrastructure components within its vicinity.

FIGURE C.13 Modeled front and rear lining during Los Angeles City Traffic vs. max regeneration power and control type [20].

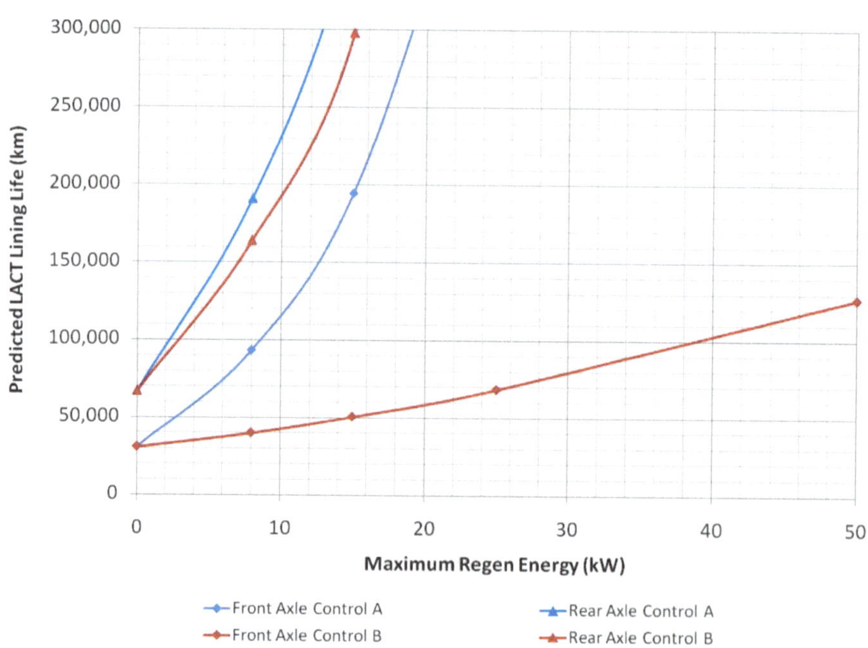

In a cooperative regenerative braking system, vehicles equipped with this technology can communicate with each other or with infrastructure such as traffic lights, road sensors, or centralized traffic management systems. Vehicles can collectively optimize their regenerative braking strategies by sharing information about their speed, direction, and braking intentions.

When one vehicle initiates braking, it can send signals to nearby vehicles or infrastructure elements, enabling them to anticipate the deceleration event. This early warning allows neighboring vehicles to adjust their speed or engage regenerative braking preemptively, maximizing energy recapture and enhancing overall efficiency.

Copper Losses

Copper losses, also known as copper resistance or resistive losses, refer to the energy dissipated in heat when electric current flows through the copper conductors of electrical components, such as motors, transformers, and wiring, in an EV system. These losses occur due to the inherent resistance of copper, which generates heat in proportion to the square of the current passing through the conductor (I^2) and the resistance of the conductor (R). Copper losses contribute to energy inefficiency in the electrical system of an EV, leading to reduced overall efficiency and increased operating temperatures. Minimizing copper losses is essential for optimizing EV components' performance and efficiency and maximizing EVs' driving range:

$$P_{\text{Cu}} = I^2 \times R$$

where

P_{Cu} is the copper loss in watts (W)
I is the current flowing through the conductor in amperes (A)
R is the resistance of the conductor in ohms (Ω)

Wire harness resistance calculation:

$$R = \rho\left(\frac{L}{A}\right)$$

where

R is the resistance of the wire harness in ohms (Ω)
ρ is the resistivity of the wire material in ohm-meters (Ω·m)
L is the length of the wire in meters (m)
A is the cross-sectional area of the wire in square millimeters (m^2)

Co-Routed Wires

Co-routed wires refer to a design approach used in EVs. Multiple electrical wires or cables are bundled within the vehicle's wiring harness and routed along the same path or conduit. This design technique aims to optimize space utilization, reduce weight, and simplify assembly by minimizing the number of separate wires runs and connectors needed in the vehicle's electrical system [13].

Co-routed cables can help minimize EMI, reduce signal crosstalk, and improve overall electrical system efficiency by bundling multiple wires and routing them along the same path [14]. Additionally, co-routing wires can enhance the reliability and durability of the wiring harness by decreasing the likelihood of wire chafing, abrasion, and damage during vehicle operation.

Co-routed wires are commonly used in EVs to streamline the wiring harness design, improve manufacturability, and enhance overall vehicle performance and reliability. This design approach contributes to the efficient and effective integration of electrical systems in EVs, supporting their transition to a cleaner and more sustainable mode of transportation.

Core Pins for AC and DC Charging

Core pins in the context of EV and HV charging are crucial components of the electrical connectors used in both AC and DC charging systems. These pins are part of the plug and socket assembly that facilitates electrical power transfer from a charging station to the vehicle's battery system.

Core pins' primary purpose is to establish a secure and efficient connection between the charging cable and the vehicle's charging port. They also ensure the correct transmission of power and communication signals required to manage the charging process.

Types:

1. **AC Charging Core Pins:**
   - **Function:** It facilitates AC transmission from the power grid to the vehicle. AC charging is typically slower and is commonly used for home and public charging stations.
   - **Configuration:** It often involves multiple pins, including phases, neutral, and ground, to handle the AC power safely.
2. **DC Charging Core Pins:**
   - **Function:** It allows DC transmission from the charging station to the vehicle, enabling much faster charging times. DC charging is ideal for rapid charging needs during long trips.

- **Configuration:** It includes positive and negative pins for the DC and additional pins for communication and safety purposes.

Materials: Core pins are typically made from highly conductive metals such as copper or brass, often plated with silver or another corrosion-resistant material to enhance conductivity and durability.

Core Pin Level

Core pins are vital in achieving the desired control plan level within EV battery manufacturing. Ensuring accurate cell placement and alignment, they contribute to meeting quality standards, performance targets, and safety regulations established for EVs. Properly defined control pilot levels associated with core pin usage help manufacturers maintain consistency, reliability, and traceability throughout production, enhancing the EV battery packs' overall integrity and functionality.

Control pilot level detection, short for charge point level detection, is a vital feature in EVs designed to facilitate seamless charging experiences. It involves the vehicle's ability to accurately detect and accurately identify the available charging infrastructure's power level or charge point level. This feature ensures compatibility between the EV's charging capabilities and the charging station, optimizing charging efficiency and safety.

Cost per kWh (Kilowatt-Hour)

Cost per kWh refers to the price charged to EV owners for the electricity consumed during charging. It measures the cost efficiency of EV operation and represents the money required to charge the EV's battery with a certain amount of energy. Cost per kWh varies depending on factors such as electricity rates, charging infrastructure, and charging location (e.g., residential, commercial, and public charging stations). Understanding the cost per kWh is essential for EV owners to estimate the expenses associated with charging their vehicles and to compare the cost of electric driving with that of conventional gasoline-powered vehicles.

Coulometric Efficiency (of Charge–Discharge Losses)

Coulometric efficiency refers to the ratio of the actual amount of charge or discharge during a specific electrochemical process in a battery to the theoretically calculated amount of charge or discharge. In EVs, coulometric efficiency measures how effectively a battery converts electrical energy into chemical energy during charging and vice versa during discharging. It accounts for losses due to inefficiencies in the charging and discharging processes, such as internal resistance, side reactions, and heat generation.

$$\text{Coulometric efficiency}\,(\eta) = \frac{Q_{Discharge}}{Q_{Charge}} \times 100\%$$

where

$Q_{Discharge}$ is the total charge delivered by the battery during discharge (in coulombs or ampere-hours)

Q_{Charge} is the total charge put into the battery during the charging process (in coulombs or ampere-hours)

Coulometric efficiency is expressed as a percentage and provides insight into a battery system's overall performance and efficiency. A higher coulometric efficiency indicates that the battery can store and deliver energy more effectively, resulting in reduced energy losses and improved battery performance. Understanding coulometric efficiency is essential for evaluating EV batteries' energy efficiency and long-term durability, and optimizing charging and discharging strategies to maximize battery lifespan and performance.

Coupler Disconnect Current Limit

The coupler disconnect current limit is a safety feature implemented in EV charging systems to protect against potential hazards during charging (Figure C.14). This feature sets a maximum threshold for electrical current that can flow through the charging coupler (connector) when disconnected from the EV charging port [21]. Below is the general equation for voltage generated in an inductive system upon abrupt termination of the current through the system:

$$V = L\left(\frac{dI}{dt}\right)$$

The coupler disconnect current limit prevents arcing or electrical sparks that may occur during the disconnection process, particularly in high-voltage charging systems. Arcing or sparking can pose safety risks, such as electrical shock or fire, especially when flammable materials or gases are present.

See also **Interlock**.

The coupler disconnect current limit helps minimize the risk of arcing and ensures EV charging equipment's safe and reliable operation by limiting the current flow during disconnection. This safety feature is typically implemented in compliance with relevant safety standards and regulations to protect users, vehicles, and surrounding environments during the charging process.

FIGURE C.14 A block diagram of the connector charging interface [21].

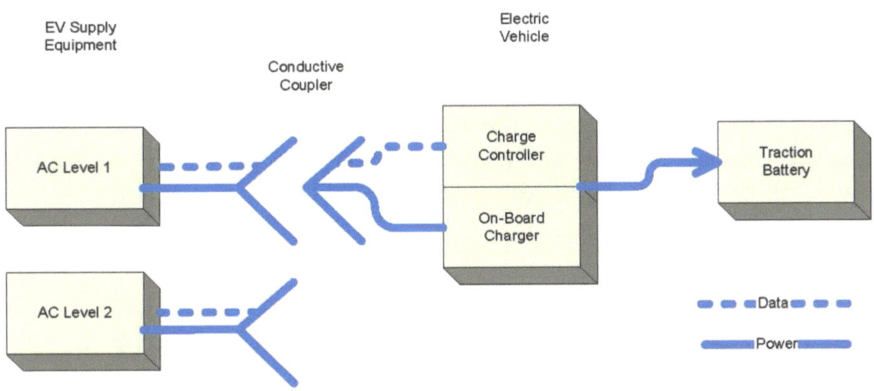

CPO (Charge Point Operator)

A CPO is an entity or organization that owns, operates, and manages EV charging stations. The CPO is responsible for installing, maintaining, and managing the charging infrastructure, including the physical charging equipment, network connectivity, and payment processing systems. CPOs may offer charging services to EV drivers through various business models, such as subscription-based memberships, pay-per-use fees, or partnerships with businesses and municipalities. Additionally, CPOs play a crucial role in expanding and improving EV charging infrastructure, facilitating the widespread adoption of EVs by providing convenient and reliable charging solutions for EV owners.

CPP (Critical Peak Pricing)

CPP is a dynamic pricing strategy utilized in EV charging infrastructure to incentivize users to shift their charging activities to off-peak periods. This pricing mechanism aims to manage electricity demand during peak periods, typically when the grid is under stress or when electricity generation from renewable sources is limited.

Many electric utilities are exploring the implementation of CPP to more effectively manage electricity demand, incentivize EV owners to charge their vehicles during off-peak hours, and support grid stability.

CPUC (California Public Utilities Commission)

The CPUC is the regulatory agency overseeing the utilities and services provided by California's investor-owned electric, gas, water, and telecommunications companies. Established by the California Constitution, the CPUC regulates the rates, safety, and quality of service these utility companies provide to ensure fair and equitable treatment of consumers while promoting sustainable and reliable energy infrastructure development. For EVs, the CPUC plays a significant role in shaping policies and regulations related to EV charging infrastructure, electric grid integration, and utility programs to support the adoption and integration of EVs into California's transportation sector. The CPUC's initiatives aim to accelerate the transition to cleaner and more sustainable transportation options while addressing the state's climate goals and reducing greenhouse gas (GHG) emissions [22].

CPV (Concentrated Photovoltaic)

CPV is a solar power generation technology that utilizes optical systems to focus sunlight onto high-efficiency PV cells (Figure C.15). Unlike traditional flat-plate PV systems, which capture sunlight over a large area, CPV systems use lenses or mirrors to focus sunlight onto small, multijunction solar cells. By concentrating sunlight, CPV systems can achieve higher solar energy conversion efficiencies and generate more electricity per unit of surface area. This technology suits locations with high solar irradiance and limited land space. CPV systems are used in various applications, including utility-scale solar power plants, remote power generation, and solar-powered EV charging stations.

FIGURE C.15 Solar energy generates power in the PV used to charge the vehicle.

Tom Kuest - Fotograf/Shutterstock.com.

Cross-Rail Connection

A cross-rail connection in an EV refers to a structural or electrical connection that enables the transfer of power or signals between different components or subsystems within the vehicle. This connection typically involves a physical interface, such as a connector or wiring harness, that facilitates the transmission of electrical currents, data signals, or control signals across various modules or systems within the EV. Cross-rail connections play a crucial role in integrating and coordinating the operation of different vehicle components, such as battery packs, power electronics, electric motors, and auxiliary systems, ensuring proper functionality and performance of the vehicle's electric drivetrain and other subsystems. These connections may be designed to withstand high currents, voltages, and environmental conditions encountered in the automotive environment while also providing reliability, durability, and ease of assembly and maintenance.

Cryptographic Modules for EVs

Cryptographic modules for EVs refer to hardware or software components that implement cryptographic algorithms and security protocols to protect sensitive data and communication within the vehicle's onboard systems (Table C.2). These modules are designed to ensure the confidentiality, integrity, and authenticity of data exchanged between various components of the EV ecosystem, including vehicle control systems, charging infrastructure, and connected services [23].

Cryptographic modules for EVs may perform encrypting communication channels between the EV's onboard systems, such as the BMS, and external entities, such as charging stations or remote servers, to prevent eavesdropping and data tampering. The module can verify the identity of entities involved in communication exchanges, such as authenticating the EV to the charging station or validating software updates received by the vehicle to prevent unauthorized access or malicious attacks [24].

Through encryption and access control mechanisms, the cryptographic module protects sensitive data stored within the EV's onboard systems, such as user credentials, vehicle configuration settings, and diagnostic information. This includes managing cryptographic keys used for encryption, decryption, and authentication purposes, including key generation, storage, distribution, and revocation, to ensure secure cryptographic operations.

TABLE C.2 An example of the various levels of expectation upon cryptographic modules [25].

	Security level 1	Security level 2	Security level 3	Security level 4
Crypto module	Specification of cryptographic module and cryptographic boundary. Description of cryptographic module including all hardware, software, and firmware components. Statement of module security policy.			
Module interfaces	Required and optional interfaces. Specification of all interfaces and of all internal data paths.		Data ports for critical security parameters physically separated from other data ports.	
Roles and services	Logical separation of required and optional roles and services.	Role-based operator authentication.	Identity-based operator authentication.	
Finite state machine	Specification of finite state machine model. Required states and optional states. State transition diagram and specification of state transitions.			
Physical security	Production grade equipment.	Locks or tamper evidence.	Tamper detection and response for covers and doors.	Tamper detection and response envelope.
EFP/EFT	No requirements.			Temperature and voltage.
Software security	Specification of software design. Relate software to finite state machine model.		High-level language implementation.	Formal model. Pre- and postconditions.
Operating system security	Executable code. Authenticated Single user, single process.	Controlled access protection (C2 or equiv.).	Labelled protection (B1 or equiv.). Trusted communications path.	Structured protection (B2 or equiv.).
Key management	FIPS approved generation/distribution techniques.		Entry/exit of keys in encrypted form or direct entry/exit with split knowledge procedures.	
Crypto algorithms	FIPS approved cryptographic algorithms for protecting unclassified information.			
EMI/EMC	FCC Part 15. Subpart J, Class A (Business use). Applicable FCC requirements (for voice).		FCC Part 15. Subpart J, Class B (Home use).	
Self-tests	Power-up tests and conditional tests.			

Courtesy of NIST.

Cryptographic modules verify the integrity and authenticity of software updates and patches delivered to the EV's onboard systems to prevent malware injection and unauthorized modifications that could compromise vehicle security and safety.

Current

See **Ampere**.

REFERENCES

[1]. Z. Wu, N. N. Manne, J. Harper, B. Chen and D. Dobrzynski, "A Cloud-Based Simulation and Testing Framework for Large-Scale EV Charging Energy Management and Charging Control," *SAE Int. J. Adv. & Curr. Prac. in Mobility*, vol. 4, no. 5, pp. 1492-1500, 2022, https://doi.org/10.4271/2022-01-0169.

[2]. Society of Automotive Engineers, "J1772 SAE Electric Vehicle and Plug-In Hybrid Electric Vehicle Conductive Charge Coupler," SAE Publishing, Warrendale, 2024.

[3]. Society of Automotive Engineers, "J1715 Hybrid Electric Vehicle (HEV) and Electric Vehicle (EV) Terminology," SAE Publishing, Warrendale, 2022.

[4]. V. Goyal, A. Dudekula and J. Naber, *Development and Validation of Dynamic Programming Based Eco Approach and Departure Algorithm*, Warrendale: SAE Publishing, 2024.

[5]. CHAdeMO EU Secretariat, "CHAdeMO," April 2019, accessed June 28, 2024, https://www.chademo.com.

[6]. Society of Automotive Engineers, "J3105 Electric Vehicle Power Transfer System Using Conductive Automated Connection Devices," SAE Publishing, Warrendale, 2023.

[7]. Society of Automotive Engineers, "J1773 SAE Electric Vehicle Inductively Coupled Charging," SAE Publishing, Warrendale, 2014.

[8]. Society of Automotive Engineers, "J2836 Use Cases for Diagnostic Communication for Plug-In Electric Vehicles," SAE Publishing, Warrendale, 2021.

[9]. Society of Automotive Engineers, "J3072 Interconnection Requirements for Onboard, Grid Support Inverter Systems," SAE Publishing, Warrendale, 2021.

[10]. Society of Automotive Engineers, "J1113-1 Electromagnetic Compatibility Measurement Procedures and Limits for Components of Vehicles, Boats (Up to 15 m), and Machines (Except Aircraft) (16.6 Hz to 18 GHz)," SAE Publishing, Warrendale, 2013.

[11]. Society of Automotive Engineers, "J2990 Hybrid and Electric Vehicle Safety Systems Information Report," SAE Publishing, Warrendale, 2020.

[12]. US Department of Transportation, *Lithium Battery Guide for Shippers*, Washington, DC: National Highway Traffic Safety Administration, 2021.

[13]. Society of Automotive Engineers, "J1742 Connections for High Voltage On-Board Vehicle Electrical Wiring Harnesses - Test Methods and General Performance Requirements," Society of Automotive Engineers Publishing, Warrendale, 2022.

[14]. Society of Automotive Engineers, "J1113-12 Electrical Interference by Conduction and Coupling - Capacitive and Inductive Coupling via Lines Other than Supply Lines," SAE Publishing, Warrendale, 2022.

[15]. Society of Automotive Engineers, "J1962 Diagnostic Connector," SAE Publishing, Warrendale, 2016.

[16]. Society of Automotive Engineers, "J1979 E/E Diagnostic Test Modes: OBDonUDS," SAE Publishing, Warrendale, 2021.

[17]. Society of Automotive Engineers, "J1699/3 Vehicle OBD II Compliance Test Cases," SAE Publishing, Warrendale, 2021.

[18]. J. Jing, Y. Liu, Y. Zhong, W. Huang and Z. Dai, *Research on the Power-On and Power-Off Control and Propulsion Control Method in Hybrid and Electric Vehicles*, SAE Publishing, Warrendale, 2024.

[19]. J. Gramling, *Fully Integrated IVT-Regenerative Braking*, Warrendale: SAE Publishing, 2014.

[20]. D. B. Antanaitis, *Effect of Regenerative Braking on Foundation Brake Performance*, Warrendale: SAE Publishing, 2010.

[21]. Society of Automotive Engineers, "J2953-1 Test Procedures for the Plug-In Electric Vehicle (PEV) Interoperability with Electric Vehicle Supply Equipment (EVSE)," SAE Publishing, Warrendale, 2023.

[22]. Lipman, T., Harrington, A., and Langton, A., "Total Charge Management of Electric Vehicles," Energy Research and Development Division, UC Berkeley, 2021.

[23]. US Government, "Federal Information Processing Standards Publication 140-3: Security Requirements for Cryptographic Modules," US Department of Commerce, Washington, DC, 2019.

[24]. National Institute of Standards and Technology, "Federal Information Processing Standards Publication 140-3: Security Requirements for Cryptographic Modules," US Department of Commerce, Washington, DC, 2019.

[25]. ISO/SAE, "Road Vehicles—Cybersecurity Engineering," International Organization for Standardization & SAE International, Warrendale, 2021.

D

DC Charging–Discharging

DC charging, often referred to as "fast charging," involves the transfer of electricity through DC to charge the batteries of EVs and HVs rapidly [1]. Unlike AC charging, which requires a converter to change AC from the power grid into DC suitable for vehicle batteries, DC chargers supply power directly in a form that can be stored in the vehicle's battery, facilitating much quicker charging times.

In DC charging, electricity flows directly into the vehicle's battery at a high voltage, significantly speeding up the charging process. This method bypasses the vehicle's OBC, allowing for higher power levels and faster charging. DC discharging, on the other hand, refers to the process where the vehicle's battery supplies DC power back to the grid or other systems. This practice is less common but growing in interest due to its potential in smart grid applications and as a backup power source (Figure D.1).

FIGURE D.1 An example of a schematic of a discharge subsystem [1].

DC–DC Converter

A DC–DC converter is an electronic device used in EVs and HVs to convert DC from one voltage level to another. This component is crucial for managing energy flow within the vehicle's electrical system, ensuring that various components receive the correct voltage.

In EVs and HVs, the DC–DC converter primarily functions to step down the high voltage from the main traction battery pack to a lower voltage necessary to power the vehicle's auxiliary systems such as lights, dashboards, infotainment systems, and HVAC (Figure D.2). This is essential because these auxiliary systems typically operate at standard automotive voltages (e.g., 12 or 24 V), whereas the traction battery operates at much higher voltages (e.g., 300 to 800 V) [2].

FIGURE D.2 (a) Conventional EV powertrain and (b) powertrain incorporating a synchronous DC–DC boost converter using insulated gate bipolar transistors (IGBTs) [2].

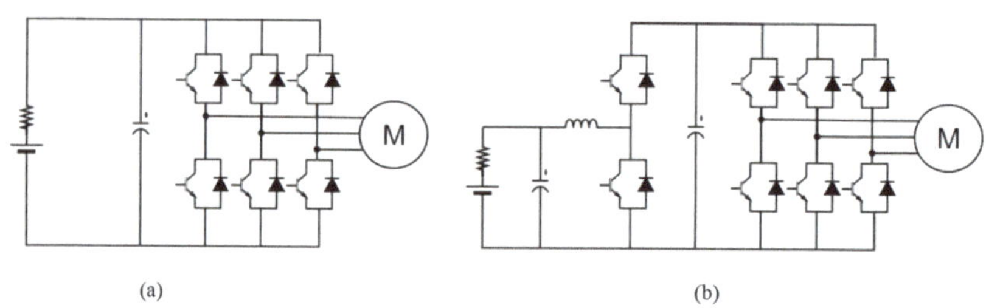

Types: There are two main types of DC–DC converters used in EVs:

1. **Buck Converter:** This type steps down the voltage from a higher to a lower level.
2. **Boost Converter:** Conversely, this type steps up the voltage from a lower level to a higher level, though it is less commonly used for auxiliary power in vehicles.

DC–DC converters' efficiency and reliability are vital for overall energy management in EVs and HVs. They help maximize the use of the battery's stored energy and ensure safety by maintaining stable voltage levels across all electronic systems.

Recent advancements in DC–DC technology focus on improving efficiency, reducing size and weight, and enhancing the ability to handle higher power levels. Innovations often include wide bandgap semiconductors, such as silicon carbide (SiC) and gallium nitride (GaN), which offer superior performance characteristics at high voltages and temperatures compared to traditional silicon-based solutions.

DC Fast Charge Alternative

DC fast charge, or DC fast charging, is a rapid charging technology for EVs and PHEVs. It utilizes DC to deliver a high electricity rate directly to the vehicle's battery, significantly speeding up the charging process compared to standard AC charging methods.

DC Fast Charging stations typically operate at 400 to 800 V or higher voltages, allowing them to deliver power levels from 50 to 350 kW and beyond. This capability enables them to charge an EV's battery to 80% in approximately 20 to 30 min, depending on the vehicle's battery size and charging capacity.

DC Fast Charging is ideal for long-distance travel where time is a constraint, and rapid charging is necessary to continue the journey. These stations are commonly found along highways, in commercial areas, and near amenities such as restaurants and shopping centers, allowing drivers to engage in activities while their vehicle charges.

Deenergized

This term refers to the state of an EV or HV when it is not connected to any source of electrical power and all systems capable of conducting electricity are inactive. It is primarily used in the context of safety and maintenance.

EVs and HVs must be entirely deenergized before performing any repairs, maintenance, or diagnostic procedures. This involves disconnecting the battery, turning off the ignition, and ensuring that all capacitors within the electronic systems are discharged.

Deenergizing a vehicle is a critical safety measure to prevent electric shock, accidental activation of the electric drive system, or other hazards associated with the high-voltage systems in these vehicles. It is a standard procedure followed in automotive workshops and by emergency responders to manage EV and HV incidents safely.

For additional information, see **J1742**, **J2840**, **J1673**, and **J2183** [3–6].

Delta Load

Delta load configuration refers to a specific arrangement of electrical components in EVs and HVs, particularly the vehicle's BMS and power distribution network. This configuration employs a delta, or triangular, connection scheme to balance the load across the battery packs, enhance stability, and improve the efficiency of power delivery.

In the delta load configuration, the battery cells or modules are interconnected in a triangular pattern, akin to the delta connection used in electrical engineering for three-phase power systems. This setup allows each system component to share a common load equally, reducing the stress on any single cell or module and thus optimizing overall battery performance and lifespan.

While the delta load configuration offers significant advantages, it also poses challenges in terms of complexity in wiring, monitoring, and maintenance. The interconnection of battery modules in a delta pattern requires sophisticated BMS software to accurately monitor and control the state of each cell or module within the network.

DER (Distributed Energy Resource)

A DER refers to a variety of small, modular power generation and storage technologies that are generally located close to where electricity is used (e.g., residential, commercial, and industrial locations). In the context of EVs and HVs, DERs can include technologies such as solar panels, small wind turbines, and battery storage systems that provide power directly to the vehicles or to the charging infrastructure supporting these vehicles.

DERs play a crucial role in enhancing the energy efficiency and sustainability of EV and HV ecosystems. By generating electricity locally, DERs can reduce dependence on traditional, centralized power grids and mitigate transmission losses. EVs equipped with bi-directional charging capabilities can act as mobile DERs by supplying stored electricity back to the grid or homes during peak demand times, a practice known as V2G or V2H.

See also, **DG (Distributed Generation)**.

Desaturation

Desaturation is a phenomenon that occurs in Li-ion batteries, commonly used in EVs, where the battery's capacity is temporarily reduced due to overdischarging or prolonged exposure to deficient SOC (Figure D.3). During desaturation, the lithium ions within the battery's electrodes become immobilized or "stuck," decreasing the battery's adequate capacity and performance. This phenomenon can occur when the battery is discharged beyond its recommended lower SOC limit or left in a deeply discharged state for an extended period. Desaturation is reversible to some extent through battery conditioning or cycling, but prolonged exposure to low SOC conditions can lead to permanent capacity loss and decreased battery lifespan. Effective battery management and charging practices are essential for minimizing desaturation and maintaining EV batteries' long-term health and performance.

FIGURE D.3 Batteries are subjected to a range of stimuli that impact performance [7].

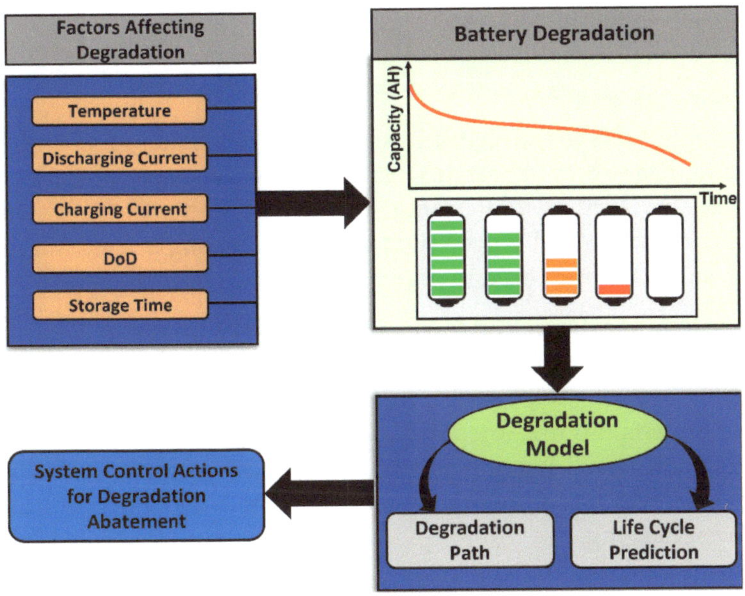

Courtesy of Laxman Timilsina.

Destination Charging

Destination charging refers to EV charging infrastructure that provides convenient and reliable charging solutions at hotels, restaurants, shopping centers, tourist attractions, and other public or private locations where EV owners spend extended periods. Unlike fast charging stations, charging stations offer slower but more accessible charging options for EV users. At the same time, they are parked at their destinations [8].

Destination charging stations typically offer Level 2 charging, which provides moderate charging speeds compared to Level 3 DC fast chargers [9]. Level 2 chargers use AC and can add around 10 to 30 mi of range per hour of charging, depending on the vehicle and charger specifications.

Destination charging stations are strategically placed at destinations where EV owners are likely to park for extended periods, such as hotels, restaurants, shopping centers, entertainment venues, workplaces, and tourist attractions [10].

Many destination charging locations offer additional amenities and services for EV owners, such as reserved parking spots, complimentary patron charging, access to nearby facilities, and integration with mobile apps for payment and monitoring.

Destination charging aims to provide EV owners convenient access to charging infrastructure during their daily routines or leisure activities. By reducing range anxiety and enhancing convenience, destination charging encourages the adoption of EVs.

Detection Logic

Detection logic refers to the algorithms and processes used in EVs and HVs to monitor and diagnose system status, faults, and efficiency parameters (Figure D.4). This system is crucial for the safe and efficient operation of the car, ensuring optimal performance and longevity of the vehicle's components, such as the battery pack, electric motors, and power electronics [11].

1. **Functionality:** Detection logic in EVs and HVs primarily identifies and responds to various operational states and potential issues. It uses sensors and software algorithms to analyze data such as voltage, current, temperature, and mechanical stresses. This information is crucial for preventing damage to sensitive components like Li-ion batteries, which require precise management to avoid overheating or overcharging.
2. **Components Monitored:** Key components monitored include the BMS, inverter, electric motors, and charging system. The detection logic ensures that these components work within their designated operational thresholds.
3. **Fault Detection and Diagnostics:** The system is adept at early detection of potential faults, significantly reducing the risk of component failure and enhancing vehicle reliability. For example, it can detect abnormalities in battery cell voltages or temperatures, triggering protective measures or maintenance alerts.
4. **Adaptive Responses:** Detection logic also involves adaptive responses to varying driving conditions and environments. This adaptability ensures the vehicle maintains optimal energy efficiency and performance, adjusting power output and regenerative braking settings based on real-time data.
5. **Software Updates:** As part of ongoing vehicle maintenance, detection logic software can be updated to improve diagnostic capabilities and incorporate new algorithms for better performance and reliability. These updates can often be performed remotely, like a smartphone or computer software update.

Detection logic is employed to diagnose faults or malfunctions within the vehicle's electrical, mechanical, or electronic systems. By analyzing DTCs and sensor data, detection logic can identify the root cause of issues and trigger warning messages or illuminate dashboard indicators to alert the driver.

FIGURE D.4 Diagnostic trouble code summary [11].

No.	Type	Description
1.	Use Case Title	Conflict and resolution
2.	Use Case Element/Scenario ID	U9 Scenario A
3.	Objectives	Customer preferences for charging may conflict with premise or facility energy management system load control parameters. Resolutions are determined either manually (business model or pricing options) or based on prescribed automated prioritization with notification.
4.	Description	Customer or EV driver can input preferences at optional three interface points: EV, EVSE, or EMS. Primary preferences are charge immediately, charge delay start time, and time charge to complete by or time charge is needed (TCIN). Note there could be future option for setting preference to automatically charge according to utility TOU rate tariff or real time price schedules (charge only during low rate periods—economy charge mode). Customer decision based on information from EV; i.e. EV calculates estimated time charge will complete based on prescribed start time, or EV calculates estimated SOC based on prescribed charge to complete by time (TCIN). There are circumstances where the customer preferences may be in conflict with the local EMS load control parameters. Examples are: • EV set for charge immediately vs EVSE set for specific delayed start time vs EMS set for prescribed alternative start time. Device or actors are not synchronized. Charging may not initiate. • EMS prescribes delayed charging or reduced charge power based on facility overall load conditions. EMS has overriding authority to control charging loads. Customer may not receive full charge. • EVSE network (acting as EMS for PEV load management) is in a curtail load mode and initiating a stagger load protocol between multiple EVs. EVSE network has overriding authority. Customer may not receive full charge. Primary Actors: Customer, PEV, EMS Secondary Actors: OEM telematics, EVSE network provider Scenario/Conditions: • Residential: PEV and premise EMS have bi-directional interface communications. • PEV is plugged in to residence EVSE and the residential EMS specifies different start time, TCIN, or power setting than PEV. • The PEV and EMS at the residence are not synchronized. • The PEV notifies customer that preference settings are different by the EMS. • Customer option is to update/change the preferences through either the PEV or EMS interface. • Option to set automatic prioritization to PEV or EMS. • Public/commercial/workplace: PEV and facility EMS have bi-directional interface communications. PEV Driver sets preference in the PEV. • PEV determines the power request and communicates preference and charge power request to the EMS network which accepts/rejects the power request based on present bulk load limitations/conditions dictated by the facility EMS. Typically, the PEV driver will prefer charging at full power immediately within these environments. Exception may be at the workplace if the EV is to be parked and plugged in for 4 hours or more. • The EMS has the authority to manipulate charge time and power due to facility load conditions (demand charge control and/or demand response). • Interruptions to charging or rejections of the PEV power request are to be communicated to the PEV driver via the PEV customer interface. • The PEV re-calculates the estimated time required to complete charging based on the SOC at time of interruption or reduced power. • Information is displayed to customer.

No.	Type	Description
		Customers may or may not be provided the right to opt in or opt out of DR and demand charge events controlled by the local EMS. It is at the discretion of the site owner what PEV load management strategy and architecture are implemented. Basic premise is the customer is provided notification about load management events that interrupt the customer charge session or overrides the customer preferences. Resolutions in favor of the PEV owner may be limited in the non-residential environments. Site owners may choose to guarantee full charging services without susceptibility to interruptions for DR and demand charge events or other by charging a premium price to the PEV owner.
5.	Prerequisite	There is an EMS at the premise or facility. Bi-directional communications path exists between PEV and local EMS. Customer preferences are able to be communicated from the PEV to the local EMS.
6.	Requirements	Each device or actor is intelligent and has an established interface protocol EMS provides real time accept/reject responses to PEV power requests. PEV provides real time notifications to customer via smart phone APP alerts when customer preferences are rejected or interruptions to charging session are activated.
7.	Post Conditions	Customer receives proper notification/response about status of charging session with ability to resolve conflicts or avoid significant impacts to driving needs. Failure Conditions: Non-interoperable communications interface protocols. Communications failure.

DG (Distributed Generation)

DG refers to the localized production of electricity closer to the point of use, as opposed to centralized power generation at large plants distant from end users (Figure D.5). In the context of EVs and HVs, DG primarily involves the integration of renewable energy sources, such as solar panels or small wind turbines, to generate power directly for the charging infrastructure.

- **V2G Systems:** EVs and HVs equipped with V2G technology can be mobile power sources. When connected to the grid, these vehicles can deliver excess electricity stored in their batteries back to the grid, particularly during peak demand times or outages, effectively supporting the grid and promoting energy stability.
- **Charging Stations:** DG can be employed at EV charging stations by installing photovoltaic solar panels, small-scale wind turbines, or other renewable sources. This reduces the dependency on the central grid and decreases the carbon footprint associated with EV charging.
- **Home Charging Systems:** For home-based charging, DG enables vehicle owners to generate their own electricity through renewable sources such as rooftop solar panels. This self-sufficiency can significantly reduce electricity costs and increase the environmental benefits of using EVs and HVs.

FIGURE D.5 Modern energy systems consist of an array of generating sources.

petovarga/Shutterstock.com.

Dielectric Breakdown or Flashover

Dielectric breakdown or flashover in EVs and HVs refers to the failure of an insulating material used in electrical components to resist electric flow, resulting in an unintended electrical discharge. This phenomenon typically occurs when the electric field within the material becomes strong enough to ionize the insulating medium, leading to a sudden increase in electric current and potential damage to the electrical system.

Dielectric breakdown is a critical concern due to the high voltages used in these vehicles' battery packs and power systems. Components such as batteries, inverters, and high-voltage cables are designed with specific insulation materials to withstand these voltages under normal operating conditions. However, excessive stress, material degradation, or manufacturing defects can lead to dielectric breakdown.

The dielectric breakdown can result in several issues:

- **Safety Risks:** It poses significant safety risks, including fires and electric shocks, endangering vehicle occupants and service personnel.
- **Component Failure:** Critical components may fail, leading to costly repairs and vehicle downtime.
- **System Reliability:** The overall reliability of the vehicle's electrical system is compromised, potentially leading to operational failures.

Dielectric Strength

Dielectric strength is a critical property for materials used in the electrical systems of EVs and HVs. It refers to the maximum electric field a material can withstand without experiencing breakdown or failure. This property is significant for ensuring the safety, efficiency, and reliability of electrical components such as wire harnesses, connectors, and integrated circuits (ICs) in these vehicles.

- **Wire Harness:** In EVs and HVs, wire harnesses are complex assemblies of electrical wires or cables that transmit signals or electrical power. They are typically insulated to prevent electrical leaks and protect against environmental factors. The dielectric strength of the insulation material is crucial because it needs to withstand high voltages, especially in high-power applications typical in EV powertrains.
- **Connector:** Connectors join sections of circuits in a vehicle's electrical system and are vital for maintaining the integrity of the electrical signals and power being transmitted. The materials used in connectors must have a high dielectric strength to avoid arcing and electrical discharge, which could lead to malfunctions or short circuits, compromising vehicle safety and functionality.

- **ICs:** ICs in EVs and HVs control various functions, from battery management to inverter control. They are often encapsulated in materials with high dielectric strength to protect them from the high-voltage environment typical in these vehicles. Ensuring adequate dielectric strength helps prevent voltage breakdown between the conductive parts within the ICs, which could lead to failures in critical vehicle systems.

Dielectric strength is measured in volts per unit thickness (V/mil or kV/mm) and is a key factor in the design and selection of electrical components for EVs and HVs, impacting their overall safety and performance.

$$\text{Dielectric strength} = \frac{\text{Breakdown voltage}}{\text{Thickness of the material}}$$

where

Dielectric strength is measured in volts per unit thickness (e.g., kV/mm or kV/in.)

Breakdown voltage is the voltage at which the material begins to conduct (break down)

Thickness of the material is the distance between the electrodes through the insulating material

Digital Control Pilot Circuit

A digital control pilot circuit is a crucial component in EVs and HVs. It is specifically designed to facilitate communication between the vehicle's charging system and the EVSE. This circuit plays a pivotal role in ensuring that the charging process is safe, efficient, and conforms to the vehicle's specific requirements.

The digital control pilot circuit transmits digital signals between the vehicle and the EVSE [11]. These signals contain information about the charging status, the maximum current available from the charging station, and the maximum current acceptance rate of the vehicle's battery system [9]. Based on this communication, the circuit helps to adjust the charging current, monitor the connection status, and ensure that the vehicle is charged within the safe operating limits of its electrical system.

The main components of a digital control pilot circuit include a communication controller, signal converters, and interface connectors that are integrated into the vehicle's OBC. The circuit uses pulse width modulation (PWM) and, in more advanced systems, may employ more complex data communication protocols like controller area network (CAN) bus or Ethernet-based systems for more detailed and faster data transmission.

Safety is a primary concern in the design of digital control pilot circuits. The circuit includes features to detect any faults or discrepancies in the charging process, such as overcurrent, undercurrent, or connectivity issues. These safety measures help to prevent electrical hazards and damage to the vehicle's battery and charging infrastructure.

DME (Distributed Managing Entity)

A DME is an entity that monitors and manages the DER through the local DER communication interface and could be, for example, a utility company, an energy aggregator, a building energy management system, etc.

For more information, see **J1715** [12].

DMS (Distribution Management System)

A DMS is an advanced software platform used in EVs and HVs to oversee and regulate the distribution of electrical energy within the vehicle's power system. This system is crucial for managing the complex interplay between various electrical components, including the battery, electric motors, and any energy recovery systems, ensuring efficient power distribution according to driving conditions and battery status.

Functionality:

1. **Energy Optimization:** DMS intelligently allocates power between propulsion, battery charging, and auxiliary systems to maximize energy efficiency and vehicle range.
2. **Battery Management:** It continuously monitors battery status, including charge levels and health, to optimize charging strategies and prolong battery life.
3. **Regenerative Braking:** It integrates regenerative braking capabilities, which recover energy during braking and redirect it for battery recharging or immediate use, enhancing overall energy efficiency.
4. **Load Balancing:** It balances the loads from various electric components to ensure stable operation of the vehicle's electrical system, thereby preventing overloads and promoting safety.

DoD (Depth of Discharge)

DoD is a measure used primarily in the context of EVs and HVs to describe the extent to which a rechargeable battery is used relative to its total capacity. It is expressed as a percentage and indicates the portion of the battery that has been discharged compared to its overall capacity.

The DoD is critical for understanding battery life and performance in EVs and HVs. A higher DoD typically means that a more significant portion of the battery's capacity is being used before recharging, affecting the battery's lifespan and efficiency. For example, a battery with a 100% DoD has been completely discharged, using all available power, whereas a battery with a 50% DoD has used half of its charge.

$$\text{Depth discharge} = \frac{\text{Capacity used}}{\text{Total battery capacity}} \times 100\%$$

where

Capacity used is the amount of energy extracted from the battery, typically measured in kWh

Total battery capacity is the total energy the battery can store when fully charged, measured in kWh

DR (Demand Response) Event Duration

DR event duration refers to the specific length of time an EV or HV is required to alter its charging behavior in response to a DR event. Utility companies or grid operators initiate these events to stabilize the electrical grid during high energy demand periods or optimize energy usage during fluctuating supply levels.

EVs and HVs are increasingly integrated into smart grid technologies that enable two-way communication between the vehicle and the grid operator. This integration allows the grid to request temporary vehicle charging rates or schedule adjustments to help balance load and prevent grid overloads.

The duration of a DR event can vary significantly based on the grid's needs, the severity of the electricity demand peak, or the availability of alternative energy sources. Typically, these events can last a few minutes to several hours. During this period, EVs may delay the start of charging or reduce the charging rate to decrease the load on the grid.

Participation in DR events often comes with financial incentives, such as reduced electricity rates or direct payments, for EV and HV owners. It contributes to a more stable and efficient energy grid. However, it may also require them to temporarily adapt their charging habits to accommodate the grid's needs.

DRFPI (Direct Radiofrequency Power Injection)

DRFPI is a wireless charging technology designed to charge EVs without physical cables or connectors. In DRFPI systems, electrical power is transferred from a charging infrastructure to the vehicle's battery through an EMF generated by a charging pad or coil installed on the ground and a corresponding receiver coil mounted on the vehicle's underside [13]. This EMF induces an electric current in the receiver coil, which is then rectified and used to charge the vehicle's battery. DRFPI offers convenience, efficiency, and ease of use compared to traditional plug-in charging methods, as it eliminates the need for manual connection and alignment of charging cables. Additionally, DRFPI can support various charging power levels and is compatible with multiple EVs. It is a promising technology that enables seamless and efficient EV charging experiences (Figure D.6).

See also, **DWPT (Dynamic Wireless Power Transfer)**.

FIGURE D.6 Wireless power transfer (WPT) flow diagram [13].

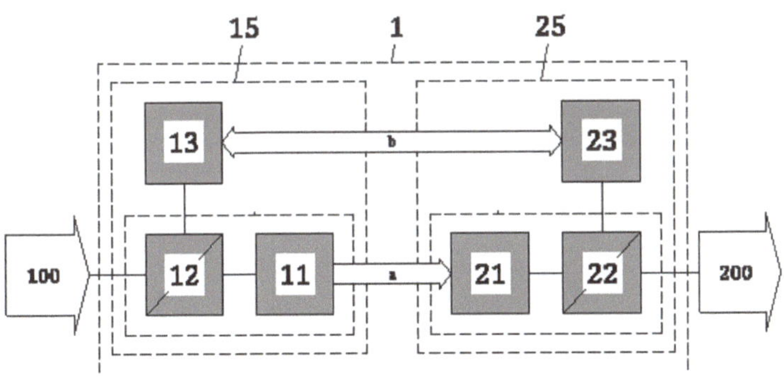

KEY			
1	WPT system	21	VA coil
11	GA coil	22	VA electronics
12	GA electronics	23	Vehicle communication controller
13	Base communication controller	25	VA - vehicle subsystem
15	GA - ground subsystem	200	Power to battery
100	Power from grid	b	Communication
a	Wireless power flow		

Drive Cycle

A drive cycle is a standardized test procedure to simulate typical driving conditions and patterns for evaluating the performance and efficiency of EVs and other vehicles. It involves a specific sequence of driving maneuvers, including acceleration, deceleration, cruising, and idling, designed to represent real-world driving scenarios.

Drive cycles vary depending on vehicle type, intended application, and regulatory requirements. Common examples of drive cycles include the Urban Dynamometer Driving Schedule (UDDS), the New European Driving Cycle (NEDC), and the Worldwide Harmonized Light Vehicles Test Procedure (WLTP).

These drive cycles aim to provide a standardized method for assessing fuel consumption, emissions, and range under controlled conditions.

- **UDDS:**
  The UDDS, or the "LA4" or "city cycle," is designed to simulate urban driving conditions with frequent stops, reflecting lower speeds and idling periods (Figure D.7). It includes a series of accelerations, steady speeds, deceleration, and idling. The cycle covers 7.5 mi and lasts 1369 sec, with an average speed of approximately 20 mph and a maximum speed of 56 mph. The UDDS is often used for vehicle emissions testing in the US [14].

FIGURE D.7 The Environmental Protection Agency (EPS) UDDS [14].

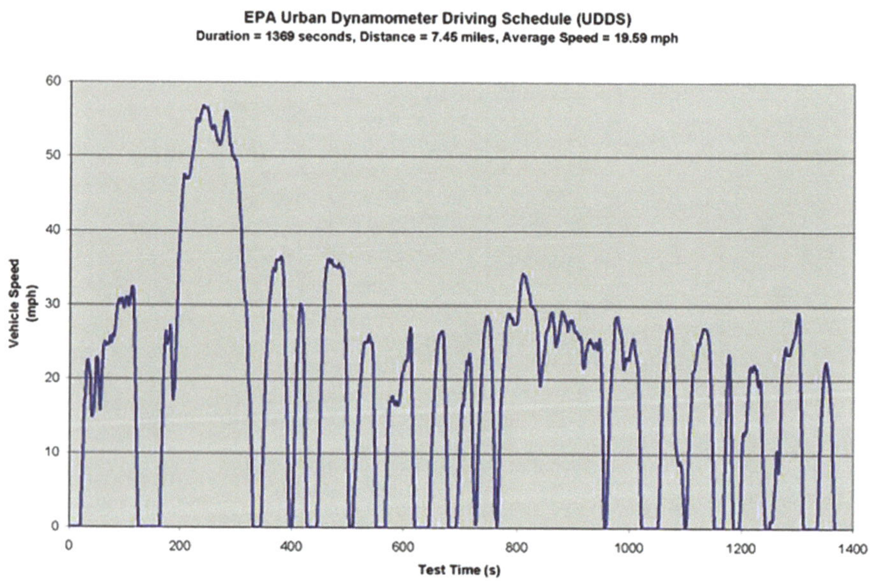

Courtesy of EPA.

- **NEDC:**
  Developed in the 1980s, the NEDC was the standard driving cycle used in the European Union (EU) for certifying the emission levels and fuel economy of light vehicles (Figure D.8). It aimed to replicate typical driving conditions in Europe and consisted of two parts: an urban cycle with low speeds and frequent stops and an extra-urban cycle with higher speeds. The total duration of the cycle was 1180 sec, covering a distance of 11 km, with an average speed of 33.6 km/h. However, it was criticized for not accurately reflecting real-world driving conditions.

FIGURE D.8 The NEDC [15].

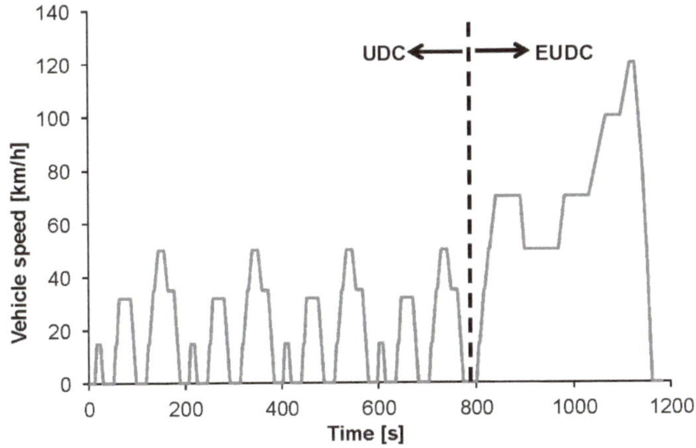

- **WLTC:**

  Introduced to replace the NEDC, the WLTC provides a more realistic test procedure that better reflects real-world driving conditions (Figure D.9). It includes four parts with different average speeds: low, medium, high, and extra high. Each part consists of various driving phases: stops, acceleration, and deceleration. WLTC uses more dynamic and realistic driving behavior to test vehicles' fuel consumption, carbon dioxide (CO_2), and pollutant emissions. The test is longer and generally tougher than the NEDC, with conditions more closely mimicking real-world driving.

FIGURE D.9 The WLTC, version 5.3, for Class 3 vehicles (≥34 kW/ton) [15].

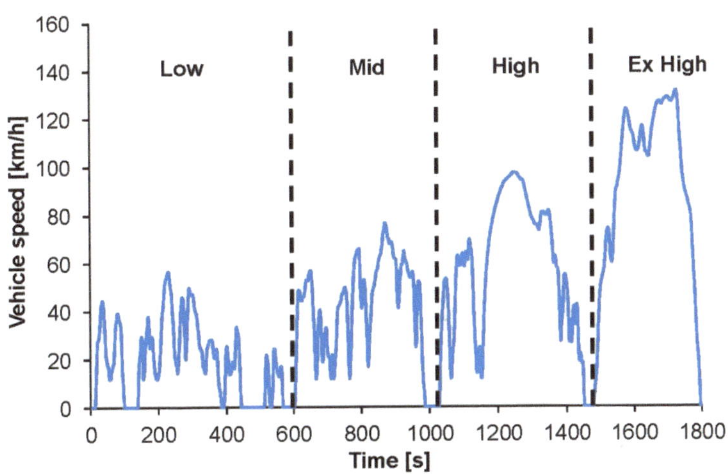

- **IDC (Indian Driving Cycle):**
  This cycle was originally developed for the Indian market to reflect typical driving conditions in India. It is relatively short and has a high average speed relative to urban congestion levels in Indian cities.
- **MIDC (Modified Indian Driving Cycle):**

  The MIDC is an evolution of the IDC, designed to better reflect real-world driving conditions by incorporating a wider range of speeds and more realistic idle conditions. In terms of its parameters, it is more closely aligned with the NEDC.

Drive cycles evaluate energy consumption, driving range, and battery performance. By subjecting EVs to standardized drive cycles, manufacturers, regulators, and researchers can compare the performance of different vehicles, assess compliance with regulatory standards, and optimize vehicle design and operation for real-world driving conditions.

Drive Profile

In the context of EVs and HEVs, a drive profile refers to a standardized or recorded set of driving conditions and behaviors that simulate how a vehicle is used in real-world scenarios. This profile includes speed, acceleration, deceleration, idle times, and road types. Drive profiles are crucial for analyzing vehicle performance, battery efficiency, and energy consumption under different driving conditions. This is used to formally evaluate the vehicle performance for the consumer, and is part of regulations.

Another definition of drive profiles allows EV drivers to customize their driving experience based on driving style, terrain, weather conditions, and the desired balance between performance and energy efficiency. Standard drive profiles may include options such as "eco mode" for maximizing energy efficiency, "sport mode" for enhanced performance, and "normal mode" for a balanced driving experience.

Drive Unit

A drive unit, also known as a powertrain or propulsion system, refers to the mechanism responsible for transmitting power from the electric motor(s) to the wheels of an EV. It encompasses components such as the electric motor(s), transmission (in some cases), differential, and associated control systems. Wheel motor configurations incorporate all elements into a single package. Such packages may be utilized at one, more, or all vehicle wheel positions.

The primary component of the drive unit, electric motors convert electrical energy stored in the vehicle's battery into mechanical energy to propel the vehicle forward. Depending on the design, EVs may feature one or more electric motors, each driving one or multiple wheels. While many EVs use a single-speed transmission due to the broad torque curve of electric motors, some models may incorporate multispeed transmissions to optimize performance and efficiency or accommodate different driving conditions.

The differential distributes power evenly to the wheels, allowing them to rotate at different speeds, particularly when cornering. In EVs, advanced electronic control systems manage torque vectoring to enhance traction, stability, and handling.

Drive units are integrated with sophisticated control systems that manage power delivery, torque distribution, regenerative braking, and other functions to optimize performance, efficiency, and safety. These control systems rely on various sensors, algorithms, and software to continuously monitor and adjust the drive unit's operation.

A single-motor rear-wheel drive (RWD) uses a single electric motor to power the rear wheels, balancing performance and efficiency. Dual-motor all-wheel drive (AWD) incorporates two electric motors—one for the front and one for the rear wheels—to deliver enhanced traction, stability, and performance, especially in challenging road conditions. Tri-motor AWD, found in high-performance EVs, such as some Tesla models, features three electric motors—two at the rear and one at the front—for acceleration and handling.

Duty Cycle

Duty cycle is connected to PWM. It refers to the fraction of time during which a signal is "on" or energized compared to the total period of the signal (Figure D.10). It is typically expressed as a percentage.

FIGURE D.10 An illustration of duty cycle of a PWM signal.

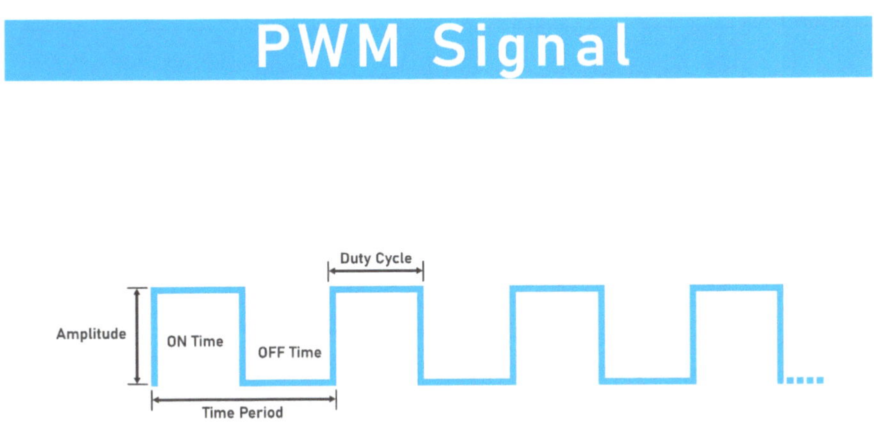

Every sketch/Shutterstock.com.

$$\text{Duty Cycle}(\%) = \left(\frac{\text{On Time}}{\text{Total Cycle Time}} \right) \times 100$$

Total Cycle Time: This is the total duration of one complete cycle of the PWM signal, consisting of both the "on" and the "off" times.

On Time: This is the duration within the cycle when the signal is active or high.

Duty Cycle (%): This is calculated by dividing the "on" time by the total cycle time and multiplying by 100 to express it as a percentage.

The duty cycle determines the average voltage delivered by the PWM signal when controlling devices like motors or light-emitting diodes (LEDs). For example, a 50% duty cycle means the signal is high for half of the total cycle time, resulting in half of the maximum possible output power being applied. Adjusting the duty cycle allows fine control over the output, which is why PWM is so effective in various applications ranging from motor speed control to brightness control in lighting systems.

DWPT (Dynamic Wireless Power Transfer)

DWPT is an innovative technology that enables the wireless transfer of electrical energy from a power source to an EV while the vehicle is in motion. Unlike traditional stationary wireless charging systems, which require the EV to be parked over a charging pad, DWPT allows continuous charging of the EV's battery pack (Figure D.11). At the same time, it moves along a roadway or designated route.

FIGURE D.11 Functional decomposition of a WPT system [16].

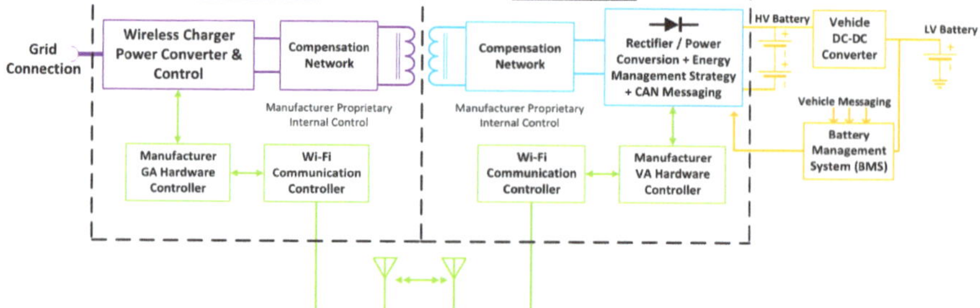

Ground assembly (GA) is the WPT subsystem that transmits energy wirelessly to a vehicle-mounted vehicle assembly (VA) [16].

This technology typically utilizes a network of coils embedded in the roadway surface, known as the wireless charging infrastructure, and a receiving coil installed on the underside of the EV. As the EV travels over the roadway, the receiving coil detects the EMF generated by the infrastructure coils, inducing an electric current in the coil and wirelessly transferring power to the EV's battery pack.

DWPT offers several advantages over traditional plug-in charging methods, including convenience, efficiency, and increased driving range for EVs (Figure D.12). By eliminating the need for manual charging and enabling continuous charging while driving, DWPT has the potential to overcome range limitations, reduce dependency on stationary charging infrastructure, and facilitate the widespread adoption of EVs.

FIGURE D.12 Dynamic wireless power transfer allows for charging while driving.

PCH.Vector/Shutterstock.com.

However, DWPT technology is still in the early stages of development and deployment, and several challenges remain, including standardization, safety, and infrastructure costs. Nonetheless, ongoing research and advancements in WPT technology hold promise for realizing the full potential of dynamic wireless charging as a critical enabler of future sustainable transportation systems.

DyFeB (Dysprosium-Iron-Boron) Permanent Magnets

DyFeB permanent magnets are advanced magnetic materials used in EVs and other applications requiring solid and compact magnets. They are part of the rare earth magnet family and are highly valued for their exceptional magnetic properties, including high magnetic strength, excellent coercivity (resistance to demagnetization), and temperature stability.

Dysprosium, an element found in small quantities within these magnets, is crucial in enhancing their magnetic properties, particularly at elevated temperatures. By incorporating dysprosium into the iron boron matrix, DyFeB magnets exhibit improved thermal stability, making them well-suited for high-temperature environments commonly encountered in EV traction motors.

DyFeB magnets in EVs contribute to the efficiency and performance of electric propulsion systems by providing magnetic solid fields for motor operation. These magnets enable higher power densities, increased torque, and improved energy efficiency, ultimately enhancing EVs' driving range and performance.

Despite their benefits, DyFeB magnets face challenges related to the scarcity and high cost of dysprosium and other rare earth elements. As a result, researchers and manufacturers are actively exploring alternative magnet materials and recycling technologies to reduce reliance on rare earth elements and ensure the sustainability of EV technologies.

Dynamic Capacity Test Discharge

Dynamic capacity test discharge is a diagnostic procedure used in EVs and HVs to assess the real-time capacity and performance of the battery pack under various operating conditions. This test involves discharging the vehicle's battery at different rates and in different environmental settings to simulate actual driving scenarios. The objective is to measure how the battery behaves in terms of energy output, efficiency, and voltage stability, which are critical for ensuring optimal performance and longevity of the battery system.

The primary purpose of conducting a dynamic capacity test discharge is to gather detailed information about the battery's capacity to deliver power under different states of charge and usage patterns. These data are crucial for the following:

- Predicting the vehicle's range under various conditions.
- Ensuring the reliability and safety of the battery system.
- Guiding maintenance and replacement schedules.
- Improving BMSs through adaptive algorithms that respond to real-time battery performance.

REFERENCES

[1]. Z. Wu, X. Su, Y. Zhu and M. Xiao, "DC Link Capacitor Active Discharge by IGBT Weak Short Circuit," *SAE Int. J. Adv. & Curr. Prac. in Mobility*, vol. 1, no. 3, pp. 1177-1187, 2019, https://doi.org/10.4271/2019-01-0606.

[2]. S. Amirpour, T. Thiringer and Y. Yu, *Mapping an Optimum DC-Link Voltage across the Entire SiC-Based EV Drive Regions Using a Synchronous Boost DC-DC Converter*, Warrendale: SAE Publishing, 2024.

[3]. Society of Automotive Engineers, "J1742 Connections for High Voltage On-Board Vehicle Electrical Wiring Harnesses - Test Methods and General Performance Requirements," Society of Automotive Engineers Publishing, Warrendale, 2022.

[4]. Society of Automotive Engineers, "J2840 High Voltage Shielded and Jacketed Cable," SAE Publishing, Warrendale, 2021.

[5]. Society of Automotive Engineers, "J1673 High Voltage Automotive Wiring Assembly Design," SAE Publishing, Warrendale, 2012.

[6]. Society of Automotive Engineers, "J2183 60 V and 600 V Single-Core Cables," SAE Publishing, Warrendale, 2012.

[7]. L. Timilsina, P. R. Badr, P. H. Hoang, G. Ozkan, B. Papari and C. S. Edrington, "Battery Degradation in Electric and Hybrid Electric Vehicles: A Survey Study," *IEEE Access*, vol. 4, pp. 1-35, 2023.

[8]. Society of Automotive Engineers, "J3068 Identification of Vehicles and Supply Equipment for Conductive AC Charging," SAE Publishing, Warrendale, 2023.

[9]. Society of Automotive Engineers, "J1772 SAE Electric Vehicle and Plug-In Hybrid Electric Vehicle Conductive Charge Coupler," SAE Publishing, Warrendale, 2024.

[10]. Society of Automotive Engineers, "J2847-2 Communication between Plug-In Vehicles and Off-Board DC Chargers," SAE Publishing, Warrendale, 2023.

[11]. Society of Automotive Engineers, "J2836 Use Cases for Diagnostic Communication for Plug-In Electric Vehicles," SAE Publishing, Warrendale, 2021.

[12]. Society of Automotive Engineers, "J1715 Hybrid Electric Vehicle (HEV) and Electric Vehicle (EV) Terminology," SAE Publishing, Warrendale, 2022.

[13]. Society of Automotive Engineers, "J2954 Wireless Power Transfer for Light-Duty Plug-In/Electric Vehicles and Alignment Methodology," SAE Publishing, Warrendale, 2022.

[14]. US Federal Government, "Emission Standards Reference Guide - EPA Urban Dynamometer Driving Schedule (UDDS)," April 18, 2024, accessed June 7, 2024, https://www.epa.gov/emission-standards-reference-guide/epa-urban-dynamometer-driving-schedule-udds.

[15]. Bielaczyc, P., Szczotka, A., and Woodburn, J., "Carbon Dioxide Emissions and Fuel Consumption from Passenger Cars Tested Over the NEDC and WLTC – An Overview and Experimental Results from Market Representative Vehicles," *2nd International Conference on the Sustainable Energy and Environmental Development*, Bielsko-Biala, Poland, 2019.

[16]. Society of Automotive Engineers, "J2847-6 Communication for Wireless Power Transfer between Light-Duty Plug-In Electric Vehicles and Wireless EV Charging Stations," SAE Publishing, Warrendale, 2020.

E

Eco Mode

Eco mode, short for "economy mode" is a driving mode feature available in many EVs that optimizes vehicle performance and efficiency to prioritize energy conservation and extend driving range. When activated, Eco mode adjusts various vehicle settings, such as throttle response, acceleration, and climate control, to minimize energy consumption and promote an eco-friendly driving experience. By moderating power output and limiting energy-intensive functions, Eco mode helps drivers maximize their EV's efficiency and reduce their environmental footprint. While Eco mode may result in slightly reduced performance compared to routine driving modes, it offers the benefit of extending driving range, which is particularly useful for long journeys or when battery range is limited. Additionally, some EVs may offer customizable Eco mode settings, allowing drivers to adjust preferences based on their driving habits and environmental priorities. Overall, Eco mode is a valuable tool for eco-conscious drivers seeking to optimize energy efficiency and minimize their environmental impact while enjoying the benefits of EV ownership.

Electric AC Compressor

See **Heat Pump**.

Electric Automobiles

Electric automobiles, commonly called EVs, utilize one or more electric motors powered by rechargeable batteries or other energy storage devices to propel the vehicle (Figure E.1). Unlike traditional ICE vehicles, which rely on gasoline or diesel fuel for propulsion, electric automobiles are powered entirely or partially by electricity. EVs come in various configurations, including BEVs, which rely solely on battery power, and PHEVs, which combine an electric drivetrain with an ICE and can operate in electric-only mode or with assistance from the engine. Electric automobiles offer several advantages over conventional vehicles, including reduced emissions, lower operating costs, and potential energy efficiency benefits. They are increasingly popular as consumers and governments seek alternatives to fossil fuel-powered transportation to mitigate environmental impacts and reduce dependence on fossil fuels.

FIGURE E.1 EVs have been around longer than you may believe.

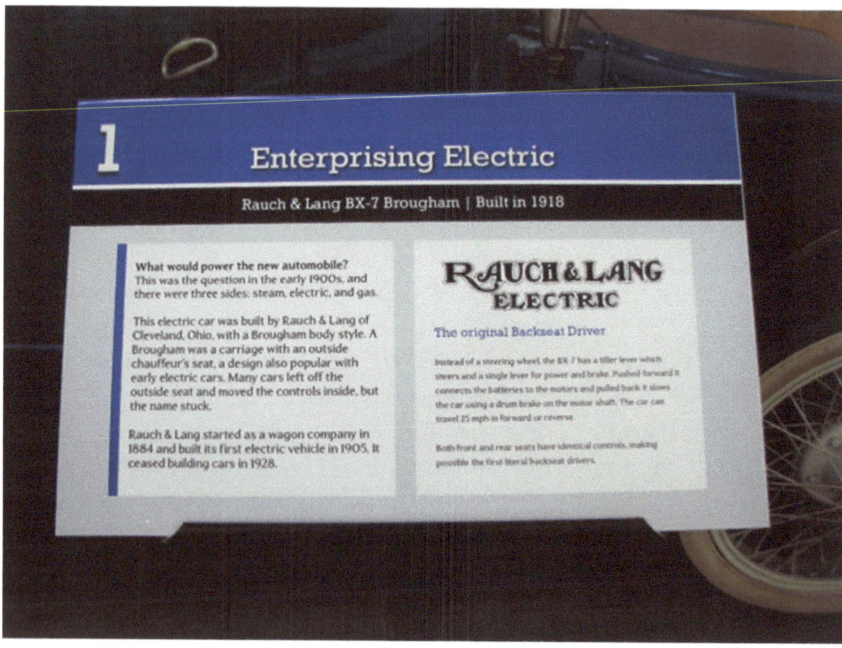

Courtesy of Jon M. Quigley.

Electric Drive Motor

An electric drive motor, commonly called an electric motor or traction motor, is a vital component of an EV propulsion system responsible for converting electrical energy from the vehicle's battery into mechanical motion to drive the vehicle's wheels. Unlike ICEs used in conventional vehicles, electric drive motors operate on the principle of electromagnetism, utilizing the interaction between electric currents and magnetic fields to generate rotational motion.

Electric drive motors come in various types, both AC and DC, with numerous variations of each, as manufacturers continue to optimize cost and performance. Notable state of the art designs seek to reduce dependence on rare earth materials and maximize reliability [1]. Each offers unique efficiency, power output, and torque delivery characteristics. Depending on the vehicle's design and configuration, these motors are typically integrated into the drive-train, directly driving the wheels or through a transmission system. Separately excited synchronous motor (SESM) is an AC induction motor made without magnets, but using slip rings or brushes to supply exciter current to the rotor.

Attributes of electric drive motors include high efficiency, instant torque delivery, and quiet operation, making them well suited for electric propulsion applications [2]. They are crucial in determining EVs' performance, efficiency, and driving dynamics, influencing acceleration, range, and energy consumption.

Electric Magnet Rotors

Electric magnet rotors are integral components of electric motors used in EVs to convert electrical energy into mechanical energy for propulsion. Unlike traditional combustion engines, which rely on combustion to generate rotational motion, electric motors utilize the principle of electromagnetism to produce torque and rotation (Figure E.2).

FIGURE E.2 A teardown of an electric motor with a highlight of the rotor.

Electric magnet rotors consist of a central shaft surrounded by a series of permanent magnets or electromagnets arranged in a specific configuration. When an electric current is passed through the coils of wire in the stator (the stationary part of the motor), it creates a magnetic field that interacts with the magnetic field produced by the rotor magnets. This interaction generates a rotating magnetic field, which induces motion in the rotor, causing it to spin.

The design and construction of electric magnet rotors are crucial in determining electric motors' performance, efficiency, and reliability. Factors such as the type of magnets used (e.g., neodymium and ferrite), the arrangement of magnets (e.g., radial and axial), and the overall design of the rotor impact the motor's torque output, speed range, and energy efficiency.

Electric magnet rotors are vital components in various electric motors used in EVs, including PMSMs (see also **PMSM**), induction motors, and switched reluctance motors (SRMs). Their efficient operation and precise control contribute to EVs' overall performance and driving experience, making them essential for electrified transportation.

Electric Motor

An electric motor is a device used in EVs and HVs to convert electrical energy into mechanical energy. This conversion is achieved by interacting the motor's magnetic field with an electric current in a wire winding to generate force in the form of torque applied on the motor's shaft. Electric motors are a key component in the propulsion system of EVs and HEVs, offering high efficiency, reliability, and quieter operation than ICEs. They can also regenerate energy through regenerative braking, improving the vehicle's efficiency.

Types:

- **AC Induction Motor:** It uses AC to produce torque and does not require permanent magnets.
- **PMSM:** This type utilizes permanent magnets on the rotor and is known for its efficiency and power density.
- **Brushless DC Motor (BLDC):** This type operates on DC with an electronic commutation system and is often used for its high power and efficiency.

Electric Power System

The electric power system of an EV refers to the integrated network of components responsible for managing the flow of electrical energy within the vehicle. This system encompasses various subsystems, including the traction battery, power electronics, electric motor(s), and associated control systems. The traction battery is the primary energy storage device, providing electrical power to drive the electric motor(s). Power electronic components, such as inverters and converters, regulate the flow of electricity between the battery and the engine, controlling speed, torque, and energy efficiency. The electric motor(s) convert electrical energy from the battery into mechanical energy to propel the vehicle. Control systems monitor and manage the operation of the electric power system, optimizing performance, efficiency, and safety. These components form the backbone of the EV's propulsion system, enabling emission-free driving and superior performance to traditional ICE vehicles. Understanding the electric power system is essential for maximizing EVs' efficiency, range, and reliability.

Electric-Powered Vehicles' Electrolyte Spillage and Electrical Shock Protection

Electrolyte spillage and electrical shock protection refers to the safety measures and technologies integrated into EVs and HVs to prevent and mitigate the risks associated with the spillage of battery electrolytes and electrical shocks. This is crucial due to the high voltage and chemical characteristics of the batteries used in these vehicles.

See **Electrolyte Spillage**.

EVs and HVs use high-voltage battery systems, typically Li-ion batteries, that pose potential hazards, including chemical burns from electrolyte leakage and electrical shocks. Effective protection mechanisms are essential to ensure the safety of occupants, emergency responders, and service personnel. The design standards and regulatory frameworks governing these vehicles are stringent, focusing on minimizing these risks through robust battery enclosures, insulation, and safety interlocks.

Electric Steering Pump

An electric steering pump is used in electric power steering (EPS) systems in EVs and some traditional ICE vehicles. Unlike hydraulic power steering systems that rely on a hydraulic pump driven by the engine, electric steering pumps utilize an electric motor to assist the steering system.

The core component of the electric steering pump is an electric motor that generates the necessary power to assist steering efforts. Electric steering pumps offer variable assistance, meaning they can adjust the level of power assistance provided to the steering system based on driving conditions, vehicle speed, and other factors. Electric steering pumps are generally more energy-efficient than traditional hydraulic power steering systems. They only operate when assistance is needed, reducing parasitic losses. Electric steering pumps are often integrated with the vehicle's ECU and other safety systems to enhance vehicle stability and responsiveness. In some EVs, electric steering pumps may be integrated with regenerative braking systems to recuperate energy during braking and improve overall energy efficiency.

Electric Traction Motor

An electric traction motor is a type of electric motor used primarily for propelling EVs, including BEVs, HEVs, and PHEVs. This motor converts electrical energy into mechanical energy to drive the vehicle's wheels, offering a cleaner, more efficient alternative to traditional ICEs.

Electric traction motors are known for their high efficiency, often exceeding 90%. They significantly reduce energy consumption and increase EVs' driving range. These motors can provide instant torque from a standstill, enabling quicker and quieter acceleration than ICE vehicles. Electric traction motors have fewer moving parts than conventional engines, require less maintenance, and are generally more reliable.

Electric Vacuum Pump

An electric vacuum pump is a crucial component in EVs designed to maintain proper braking functionality, especially when the ICE is not present to generate vacuum pressure. Unlike traditional vehicles with a vacuum pump powered by the engine's mechanical operation, EVs use an electric vacuum pump to ensure consistent and reliable braking performance.

Electrical Isolation

Electrical isolation in EVs refers to physically separating electrical components or circuits to prevent unintended electrical connections or interference. This isolation is crucial for ensuring safety, protecting sensitive electronic systems, and avoiding electrical hazards.

Electrical isolation safeguards against electric shock hazards by preventing unintended contact between high-voltage components and conductive surfaces accessible to occupants or service personnel. It also helps protect sensitive electronic components from damage caused by electrical faults, surges, or short circuits. Isolating electrical circuits facilitates fault detection and troubleshooting by containing electrical faults within isolated sections, preventing them from propagating to other parts of the vehicle's electrical system.

Electrical isolation minimizes EMI by preventing unwanted electrical coupling between different systems or components, which can affect the performance of electronic devices or communication systems.

Many regulatory standards and safety certifications require adequate electrical isolation measures to be implemented in EVs to ensure compliance with safety and performance requirements.

There are a few approaches to electrical isolation. Physical barriers, such as insulating materials or barriers, separate electrical components, or circuits, prevent direct contact or electrical coupling. Isolation transformers electrically isolate different electrical system parts, allowing power transfer and ensuring galvanic isolation between circuits. Optical isolation uses light-based signals to transmit data or control signals between isolated circuits, providing electrical isolation while allowing for system communication.

Electrical Isolation to Coolant

Electrical isolation to coolant refers to a critical safety feature in EVs that prevents electrical components from encountering the coolant system. This isolation is essential to mitigate the risk of electrical shock, or short circuits caused by the presence of coolant within the vehicle's thermal management system. By ensuring a barrier between electrical components and the coolant, EV manufacturers enhance the safety and reliability of their vehicles.

In an EV, the coolant system regulates the temperature of critical components such as the battery pack, motor, and power electronics. This system circulates coolant through channels or pipes to dissipate excess heat generated during vehicle operation. Electrical isolation of the coolant is achieved through specialized materials and design features that create a physical barrier between the coolant and electrical components. Standard methods include insulating materials, such as nonconductive plastics or coatings, and strategic placement of electrical components to minimize the risk of exposure to coolant.

Electrical isolation to coolant is subject to stringent safety standards and regulations established by automotive governing bodies and certification agencies. These standards outline requirements for insulation materials, design specifications, and testing procedures to ensure the integrity of the electrical isolation system. Compliance with these standards is mandatory for EV manufacturers seeking certification and approval for their vehicles in global markets.

Electrical Shock Protection

Electrical shock protection refers to the safety measures and technologies designed to prevent injury or death from electric shock to occupants, service personnel, and emergency responders interacting with EVs. Given the high-voltage systems typically used in EVs, ranging from 400 to over 800 V for some models, ensuring protection against electrical shock is a critical aspect of EV design and maintenance.

High-voltage components and cables are insulated to prevent accidental contact. Insulation resistance is regularly tested to ensure its integrity over the vehicle's lifetime. EVs are equipped with isolation monitoring devices that continuously check for any breakdown in the insulation barrier between the high-voltage system and the vehicle's chassis. Any compromise triggers a fault warning to prevent shock risks. Physical barriers shield high-voltage components, preventing accidental contact during routine checks or maintenance.

These components interrupt the power supply during a short circuit or overload, preventing potential electric shock incidents. High-voltage interlock loop (HVIL) is a safety feature that ensures the high-voltage system is deactivated when high-voltage connectors are disconnected or when protective covers are removed. Design standards ensure that all high-voltage connectors and conductive parts cannot be touched directly, adhering to international safety standards such as protection against access to hazardous parts with a finger (IPXXB) and protection against access with a wire (IPXXD).

Various international standards and regulations govern electrical shock protection in EVs to ensure uniform safety practices worldwide.

For additional information, see **SAE J1766**, **ISO-6469-1**, and **IEC-61851-23** [3–5].

Electrolyte

In EVs and HVs, an electrolyte is a battery component that stores electrical energy. It is a medium that allows the transfer of ions between the cathode and the anode within the battery cell, which is essential for generating electrical current.

Electrolytes can be in liquid, gel, or solid form, each serving the same fundamental purpose but offering different benefits and challenges. In EVs and HVs, electrolytes are typically found in Li-ion batteries, which are favored for their high energy density and efficiency. The composition and quality of the electrolyte significantly affect the battery's performance, safety, longevity, and overall efficiency of the vehicle (Table E.1).

TABLE E.1 An example of the content report of an analysis of fluid [6].

SAE J3042 Recommended Practice for Determining Qualitative Material Properties of Li-Ion Battery Electrolyte Report					
Section	**Parameter**	**Units**	**Qualitative desirable properties**	**Test method**	**Result**
5.1	Chemical content (impurities)	ppm or ppb	As low as possible	ASTM E260-96 ASTM D7111-16 ASTM E1184 ASTM UOP714-07 ASTM UOP389-15	
5.2	Water content	ppm	As low as possible	ASTM E1064-16	
5.3	Free acid	mg KOH/g	As low as possible	ASTM D664-17a	
5.4	Platinum-cobalt color	Pt-Co units	As low as possible	ASTM D1209-05 ASTM D5386-16	
5.5	Density/specific gravity	g/mL		ASTM D891-18 ASTM D4052-18	
5.6.1	Dynamic viscosity	mPa/s	As low as possible	ASTM D7867-13	
5.6.2	Kinematic viscosity	mm^2/s	As low as possible	ASTM D445-17a ASTM D7042-16e3	
5.7	Ionic conductivity	mS/cm	As high as possible	ASTM D1125-14	
5.8.1	Differential scanning calorimetry (DSC)	°C ΔE		ASTM E537-12	
5.8.2	Accelerating rate calorimetry (ARC)	°C/min	Self-heat rate: As low as possible	ASTM E1981-98	
5.8.3	Flammability	ppm °C	Higher concentration; higher temperature	ASTM E918-83 ASTM E681-09	
5.8.4	Flash point	°C		ASTM E502-07	
5.9	Voltage stability	V	Oxidation >4.3 V Reduction <0.1		
6	Sample disposal process			ASTM D4447-15	

Key Features:

1. **Ion Conductivity:** High ion conductivity is critical for efficient battery operation, allowing for quick charging and discharging cycles.
2. **Chemical Stability:** Electrolytes must be chemically stable under the operational temperature range and during the electrochemical reactions that occur during battery charging and discharging.
3. **Safety:** The electrolyte's chemical properties should minimize risks such as overheating, fire, or explosion, which are critical for vehicle safety.
4. **Longevity:** The electrolyte's ability to maintain its properties over many charging cycles directly influences the lifespan of the battery and, by extension, the vehicle.

Types:

- **Liquid Electrolytes:** Commonly used due to their excellent ionic conductivity. However, they can pose safety risks due to their flammability.
- **Solid Electrolytes:** Offer increased safety and are considered for next-generation solid-state batteries. They generally provide better stability but currently suffer from lower ionic conductivity compared to liquid types.
- **Gel Electrolytes:** Serve as an intermediate, offering better safety than liquid types and better conductivity than solid types.

Electrolyte Spillage

Electrolyte spillage refers to the accidental leakage or discharge of the electrolyte solution from the battery pack of an EV. The electrolyte is a liquid or gel substance that facilitates the flow of electrical current between the battery's positive and negative terminals. Typically, electrolyte spillage occurs due to damage to the battery pack, such as physical impact, overheating, or manufacturing defects, and more importantly flammability.

Impact or collision with external objects can rupture the battery casing, leading to electrolyte leakage. Excessive heat buildup within the battery pack can cause the electrolyte to expand and leak out through seals or vents. Faulty assembly or manufacturing processes may result in weak seals or compromised components, increasing the likelihood of electrolyte spillage. Over time, battery degradation and wear may weaken the integrity of the battery pack, making it more susceptible to electrolyte leakage.

Electrolyte spillage can cause corrosion of surrounding components and electrical connections, reducing performance and reliability. Depending on the composition of the electrolyte, spillage may pose environmental hazards, contaminating soil and water sources. Electrolytes are often corrosive and may be toxic or harmful upon contact with skin or inhalation of fumes, posing safety risks to individuals near the spillage.

Electromechanical Flywheels

Electromechanical flywheels are advanced ESSs in EVs that capture and store kinetic energy during braking or deceleration. Unlike traditional chemical batteries, which store energy through chemical reactions, flywheels store energy through rotational kinetic energy. These systems have a high-speed spinning rotor encased within a vacuum chamber to minimize air resistance and frictional losses. Through regenerative braking, kinetic energy from the vehicle's motion is transferred to the flywheel and stored until needed for acceleration or to power vehicle systems.

Electromechanical flywheels store energy in the form of rotational kinetic energy, offering a high power-to-weight ratio and rapid charge and discharge capabilities. These flywheels play a role in regenerative braking, turning kinetic energy to electrical energy. Electromechanical flywheels have low self-discharge rates and can efficiently capture and release energy, making them suitable for high-demand applications such as hybrid and EVs. In addition to EVs, electromechanical flywheels are used in various applications requiring high-power energy storage, including grid stabilization, uninterruptible power supplies (UPSs), and hybrid propulsion systems for buses and trains.

EMF (Electromagnetic Field)

An EMF in the context of EVs refers to the area of influence surrounding electrically charged components within the vehicle, such as the battery, motor, and power electronics. These components generate EMFs as they operate, which can extend beyond the physical boundaries of the vehicle.

The electromotive force generation in an EV application, particularly for its electric motor, can be understood through Faraday's law of electromagnetic induction. The equation can represent the EMF generated in the motor's windings:

$$\mathrm{EMF} = N \times \left(\frac{\mathrm{d}\Phi}{\mathrm{d}\tau} \right)$$

where

EMF is the electromotive force in volts (V)

N is the number of turns in the coil within the motor

$\mathrm{d}\Phi/\mathrm{d}\tau$ is the rate of change in magnetic flux through the coil in webers per second (Wb/s)

EMFs are produced by the flow of electric current through conductive materials, such as wires, coils, and electronic circuits, within the EV's powertrain. Types of EMFs: EVs primarily generate low-frequency EMFs, typically 50 to 60 Hz, associated with the frequency of the AC used in most electrical systems.

While EMFs are a natural by-product of electric systems, concerns have been raised about potential health effects associated with prolonged exposure to EMFs. However, scientific studies have not established a clear link between EMF exposure from EVs and adverse health effects.

Various regulatory bodies, such as the International Commission for Non-Ionizing Radiation Protection (ICNIRP) and the Institute of Electrical and Electronics Engineers (IEEE), have established guidelines and safety standards for EMF exposure to ensure public safety. EV manufacturers implement design measures and shielding techniques to minimize EMF exposure levels inside and outside the vehicle, ensuring compliance with safety standards and addressing public concerns.

Terminology related to EMF includes electromagnetic compatibility (EMC), which is the ability of EVs and their components to operate without causing or being affected by EMI. EMF, generated by EMI, refers to the unintended voltage that can be induced in a conductor due to the presence of fluctuating EMFs. When a conductor, such as a wire in an electronic circuit or a vehicle's wire harness, is exposed to EMI it can act as an antenna, picking up these EMIs and generating an EMF. This EMF can introduce unwanted currents into the system, leading to interference with normal operations, noise in signal processing, or even component failure if the induced voltage exceeds certain thresholds.

EMI (Electromagnetic Interference)

EMI refers to the disturbance caused by electromagnetic radiation emitted from various electrical components within an EV, which can interfere with the operation of electronic devices or systems both within the vehicle itself and in the surrounding environment.

Components such as inverters, converters, and charging systems within the EV can generate EMFs as they operate, potentially interfering with sensitive electronic devices. High-voltage battery systems in EVs can also emit electromagnetic radiation, particularly during charging or discharging cycles, which may lead to EMI. Electric traction motors can produce EMFs that may interfere with nearby electronic equipment or communication systems.

Auxiliary Systems: Various auxiliary systems in EVs, such as HVAC systems, as well as entertainment and navigation systems, can contribute to EMI.

EMI can disrupt wireless communication systems, including Bluetooth, Wi-Fi, and cellular networks, affecting connectivity within the vehicle and with external devices. Sensitive electronic devices within the EV, such as navigation systems, infotainment displays, and sensors, may experience malfunctions or erratic behavior due to interference from electromagnetic radiation. In extreme cases, EMI can affect critical safety systems in EVs, such as brake-by-wire systems, electronic stability control, and driver assistance systems, potentially compromising vehicle safety.

Compliance with EMC standards and regulations is essential to ensure that EVs do not emit excessive electromagnetic radiation that could interfere with other electronic devices or systems.

We have several solutions to mitigate the impact of EMI. Enclosures or shielding materials can contain electromagnetic radiation emitted by high-power electrical components within the vehicle, reducing the likelihood of interference with external devices. Installing filters and suppressors on power lines and communication cables can help attenuate EMI and prevent it from propagating to sensitive electronic equipment. Proper electrical components and systems grounding can mitigate EMI by providing a low-impedance path for stray currents to dissipate.

EV manufacturers conduct EMC testing to ensure that their vehicles comply with regulatory standards and do not emit excessive electromagnetic radiation that could cause interference. Physically separating sensitive electronic components from sources of electromagnetic radiation within the vehicle can reduce the risk of interference and improve overall system reliability.

EMS (Energy Management System)

An EMS in the context of EVs refers to a sophisticated electronic system designed to optimize the usage and distribution of energy within the vehicle's powertrain. This system enhances EVs' efficiency, performance, and range by intelligently managing energy flow, battery charging, and regenerative braking.

See also **EPCU (Electric Power Control Unit)**.

En Route Charging

En route charging, or on-the-go or mid-journey charging, refers to charging an EV in transit, typically during a journey between two destinations. Unlike traditional charging methods where the vehicle is stationary at a charging station, en route charging involves utilizing charging infrastructure strategically located along highways, roads, or other travel routes to replenish the vehicle's battery while it is in motion.

See also **DWPT (Dynamic Wireless Power Transfer)**.

Enclosed Pin and Socket Connection

An enclosed pin and socket connection is a type of electrical connector used in EVs and HVs for charging purposes. This connector system features a male pin and a female socket encased in a protective housing, ensuring a secure and environmentally protected connection. The design of the enclosed pin and socket connection aims to provide a robust interface for transmitting electrical power from the charging station to the vehicle. This connection type is essential for maintaining the integrity and safety of the electrical connection under various environmental conditions, especially moisture and other conducting materials [5, 7].

Energy Portal

An energy portal refers to a centralized platform or interface within an EV charging infrastructure that facilitates various functions related to energy management, monitoring, and optimization. It is a gateway for EV owners, charging network operators, utilities, and other stakeholders to interact with the charging ecosystem, access relevant data, and manage charging sessions effectively.

Energy portals allow EV owners to initiate, monitor, and manage charging sessions remotely, providing real-time visibility into charging status, battery levels, and charging rates. Energy portals facilitate seamless billing and payment processes for charging services, allowing users to view charging costs, access payment options, and receive invoices electronically.

See also **Charging**.

EPCU (Electric Power Control Unit)

The EPCU is a critical component of EVs and HVs. It manages and controls the electrical power from the battery to various vehicle systems. It is the central hub for directing electricity, ensuring efficient power distribution to the electric motor, charging, and auxiliary systems (Figure E.3).

FIGURE E.3 Typical EV and power controller [8].

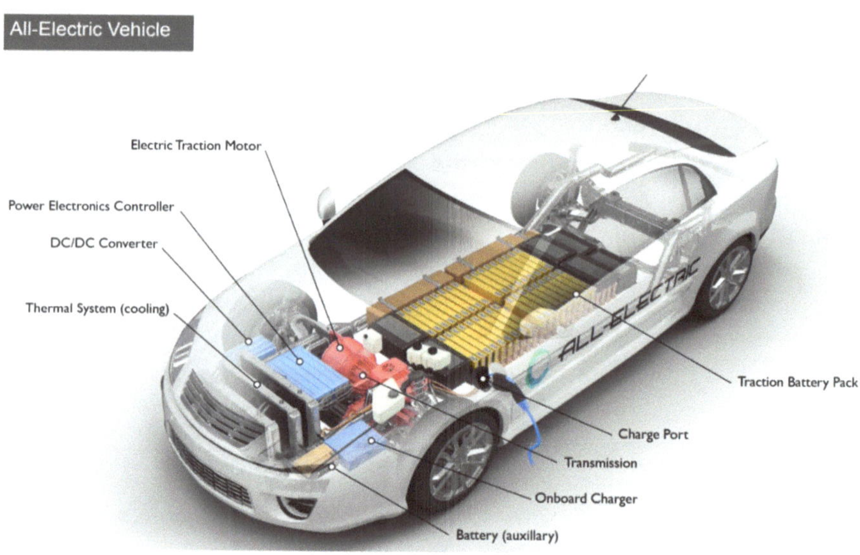

The EPCU performs several key functions:

1. **Voltage Regulation:** It converts the DC voltage from the battery to the AC voltage the electric motors need and adjusts the voltage levels depending on the power demand.
2. **Power Management:** It efficiently manages power distribution to optimize the vehicle's performance, battery life, and energy consumption.
3. **Motor Control:** It controls the speed and torque of the electric motor by adjusting the frequency and amplitude of the supplied electrical current.
4. **Thermal Management:** It works with cooling systems to manage the heat produced by electrical components, ensuring they operate within safe temperature ranges.
5. **Safety Monitoring:** It incorporates safety protocols to detect and respond to electrical faults, overcurrent situations, or other potential safety hazards.

The EPCU is essential for EVs' and HVs' performance, efficiency, and safety. It plays a pivotal role in achieving optimal energy use, enhancing driving dynamics, and extending the vehicle's range and lifespan. By precisely controlling the flow of electrical energy, the EPCU ensures that EVs and HVs are powerful, reliable, and safe to operate.

EPL (Electrical Physical Layer)

The EPL in EVs refers to the foundational infrastructure for transmitting electrical power between components within the vehicle's powertrain system. EPL encompasses the wiring, connectors, and associated components that facilitate the flow of electricity from the energy source (typically the battery) to various subsystems such as the motor, controller, and auxiliary systems.

EPL includes wiring harnesses that interconnect the different electrical components within the vehicle, ensuring the efficient transmission of power and data signals. It incorporates connectors that facilitate the secure and reliable connection between wiring harnesses and electrical components, enabling seamless integration and modularity within the vehicle's electrical system. EPL components are designed to meet specific voltage and current ratings to ensure compatibility with the vehicle's power requirements and safety standards.

EPL components may feature shielding and insulation materials to protect against EMI, moisture, temperature fluctuations, and mechanical stress, thereby ensuring the reliability and longevity of the electrical connections. Some EPL systems may include diagnostic features such as built-in sensors or communication protocols that enable real-time monitoring of electrical parameters, fault detection, and troubleshooting.

The EPL is fundamental to the operation and performance of EVs, serving as the backbone of the vehicle's electrical architecture. It plays a critical role in ensuring the safe, reliable, and efficient power transmission throughout the vehicle, impacting factors such as vehicle range, acceleration, and overall drivability. Robust EPL design and implementation are essential for optimizing EVs' performance, durability, and safety.

Electrical ratings refer to specifications that describe the electrical characteristics and capabilities of various components within the vehicle's powertrain and charging system. These ratings are crucial for understanding EVs' performance, efficiency, and compatibility with charging infrastructure.

Battery capacity denotes the energy stored in an EV's battery pack and is measured in kWh. It indicates the maximum energy the battery can deliver to power the vehicle's electric motor. Voltage is the potential difference between two points in an EV's electrical system, typically measured in volts (V). Higher voltage systems can support faster charging rates and may offer improved efficiency. Current is the measure of the flow of electric charge within the vehicle's electrical system and is measured in amperes (amps). Charging stations and OBCs specify maximum current ratings, determining the charging speed and compatibility with different power sources. Charging power measures, the rate at which energy is transferred from a charging station to the EV's battery in kW. It depends on the voltage and current levels the charging infrastructure supports and influences charging speed and efficiency.

Electrical ratings are essential for EV owners, charging infrastructure providers, and manufacturers to ensure compatibility, optimize charging performance, and effectively plan charging sessions. Compatibility between the EV's electrical system and charging infrastructure is crucial to prevent damage to components and ensure safe and efficient charging operations.

Equipment Ground (Grounding Conductor)

An equipment ground is a conductor used to connect the non-current-carrying metal parts of the EV supply equipment to the system grounded conductor, the grounding electrode conductor, or both at the service equipment.

For more information, see **J1715** [9].

EREV (Extended Range Electric Vehicle)

An EREV is a type of PHEV that incorporates an ICE along with an electric propulsion system. EREVs are designed to offer both all-electric driving capabilities and extended range through a backup gasoline or diesel generator, providing flexibility and mitigating range anxiety for drivers.

See also **HEV (Hybrid Electric Vehicle)**.

ESC (Energy Service Certificate)

An ESC is a digital or physical certificate designed to authenticate and document the amount of renewable energy used by EVs and HVs. This certificate serves as a verification tool to confirm that the energy consumed by the vehicles is sourced from renewable energy supplies, thereby supporting sustainable transportation initiatives.

The ESC is part of broader efforts to promote cleaner, greener transportation methods. By using ESCs, vehicle owners and fleet operators can certify their commitment to reducing carbon footprints and supporting renewable energy integration. This certification can be critical for compliance with environmental standards and regulations and enhance companies' brand value by showcasing their sustainability efforts.

ESD (Electrostatic Discharge)

ESD refers to the sudden flow of electricity between two electrically charged objects caused by contact, an electrical short, or dielectric breakdown. In EVs and HVs, ESD can occur during various interactions with the vehicle, such as touching the vehicle's surface, connecting charging cables, or during maintenance.

ESD is a significant concern in EVs and HVs due to the sensitive electronic components used in these vehicles, such as control units, sensors, and infotainment systems. The discharge can damage these components, potentially leading to failures or malfunctions. Vehicle manufacturers employ various ESD protection strategies to mitigate these risks and ensure the longevity and reliability of electronic systems.

For more information, see **J551** and **J1113** [10, 11].

ESS (Energy Storage System)

An ESS is a crucial component in EVs that stores electrical energy for later use. It comprises various technologies designed to efficiently capture, store efficiently, and discharge energy, enhancing EVs' performance, range, and overall efficiency.

The most common form of ESS in EVs is a high-capacity rechargeable battery pack, typically comprised of Li-ion cells. These batteries store electrical energy converted from an external power source or regenerated during braking and deceleration. Many EVs utilize regenerative braking systems, which convert kinetic energy generated during braking into electrical energy, thereby replenishing the ESS and extending the driving range.

ESS often includes power electronic components such as DC–DC converters and inverters, which manage the flow of electrical energy between the battery pack, electric motor, and other vehicle systems. ESS technology continues to advance, with ongoing efforts to increase energy density (energy storage per unit volume or weight) to extend vehicle range without significantly increasing weight or size.

See also **Battery**.

ESS (Energy Storage System) Ventilation

ESS ventilation ensures adequate airflow and ventilation around the energy storage components of an EV, such as the battery pack, to maintain optimal operating conditions and prevent overheating or thermal runaway events. As EV batteries are crucial for storing and supplying electrical energy to power the vehicle, proper ventilation is essential to dissipate heat generated during charging, discharging, and operation [12].

Effective ventilation helps dissipate heat generated by the battery pack during charging and discharging cycles, preventing temperature buildup that could degrade battery performance or lead to safety hazards. Proper ventilation is a critical aspect of an EV's overall thermal management system, which includes cooling mechanisms such as liquid cooling, air cooling, or a combination of both to regulate battery temperature within safe limits [13].

Adequate ventilation reduces the risk of thermal runaway events, which occur when battery cells overheat and release energy rapidly, potentially leading to fires or explosions. Proper ventilation helps mitigate this risk by maintaining battery temperature within safe thresholds. Optimal ventilation contributes to the overall efficiency of the ESS by ensuring that the battery operates within its optimal temperature range, maximizing energy storage capacity and extending battery lifespan.

See also **Thermal Management System**.

EUMD (Electric Vehicle Usage and Maintenance Data) Owner

An EUMD owner refers to an individual, organization, or entity with access to and is responsible for collecting, managing, and utilizing usage and maintenance data related to EVs. This role typically encompasses EV owners, fleet operators, automotive manufacturers, and third-party service providers who gather and analyze data to optimize EV performance, enhance user experience, and ensure proper maintenance.

EUMD owners are responsible for collecting a wide range of EUMD, including mileage, charging history, battery health, vehicle diagnostics, and service records. They oversee EV data storage, organization, and management, ensuring its integrity, security, and accessibility for analysis and decision-making purposes. EUMD owners analyze usage and maintenance data to gain insights into EV performance, driving patterns, charging behavior, and potential issues or anomalies that may require attention.

EUMD (Electric Vehicle Usage and Maintenance Data) Software

EUMD is software that facilitates the management and operation of EVs and HEVs. It primarily focuses on enhancing user interaction with the vehicle's systems, improving the overall driving experience, and optimizing vehicle performance.

EUMD software is integral to the user interface systems of EVs and HVs, allowing drivers to interact seamlessly with their vehicles. This software controls various aspects of the vehicle's operations, from monitoring battery status to managing energy consumption and vehicle diagnostics. It is designed to provide a user-friendly interface that delivers real-time data and actionable insights, thus enhancing vehicle usability and efficiency.

EV (Electric Vehicle)

An automotive-type vehicle, intended for public roadway use, for which propulsion is powered by an electric motor that draws primarily from a rechargeable energy storage system (RESS). An automotive-type vehicle is defined as follows:

- Any four-wheeled vehicle propelled by an ICE using on-board fuel or by an electric motor drawing current from a rechargeable storage battery or other portable energy devices.
- Which is manufactured primarily for use on public streets, roads, and highways.
- Which is rated not more than 3855.6 kg (8500 lb), which has a curb weight of not more than 2721.6 kg (6000 lb), and which has a basic frontal area of not more than 4.18 m^2 (45 ft^2).

For more information, see **J1715** [9].

EV Charge Point

See **EV/PHEV Charging System**.

EV Charging Equipment

EV charging equipment refers to the infrastructure and devices used to recharge the batteries of EVs. This equipment encompasses a range of technologies, from basic charging cables to advanced charging stations, designed to provide electric power to EVs efficiently and safely.

Charging stations, also known as EVSE, are dedicated units installed in various locations such as homes, workplaces, and public areas. They provide electric power for vehicle charging and have different configurations to accommodate multiple charging speeds and connector types.

Charging cables connect an EV to a charging station or power source. They vary in length and connector type, and some EVs are equipped with onboard charging cables for added convenience. Onboard charging equipment is located within the EV and manages the charging process. It regulates power flow, communicates with the charging station, and ensures safe and efficient charging.

EV Charging Speeds

EV charging speeds are the rate at which EVs and HEVs can recharge their batteries from an external power source. The speed at which a vehicle charges is influenced by several factors, including the charger's power output, the maximum charging rate the vehicle can accept, and the state and capacity of the vehicle's battery.

See **Level 1 Charging**, **Level 2 Charging**, and **Level 3 Charging**.

EV Conductive Coupling

EV conductive coupling refers to transferring electrical energy from a power source to an EV for charging purposes. This method involves physically connecting the EV to a charging station or power source using conductive elements, such as cables and connectors, to establish an electrical connection. Conductive coupling is commonly used in Level 1, Level 2, and DC fast charging systems, where the EV is directly connected to the charging infrastructure through a conductive pathway [7]. During the charging process, electrical energy flows through the conductive elements, allowing the EV's battery to recharge. Conductive coupling offers a reliable and efficient means of charging EVs, with charging speeds and capabilities varying depending on the charging equipment and infrastructure used [14].

EV Connector

An EV connector is an off-board component used to interface with the vehicle-mounted EV inlet to supply power and provide communication interface.

For more information, see **J1715** [9].

EV Crashworthiness

Crashworthiness in EVs refers to the vehicle's ability to protect occupants from injury and minimize damage to the vehicle's structure in the event of a collision or crash. It encompasses various design features, materials, and safety systems to enhance occupant protection and structural integrity during impact scenarios.

EV manufacturers employ advanced engineering techniques to design vehicle structures that efficiently absorb and dissipate crash forces, thereby minimizing intrusion into the passenger compartment. Crashworthiness considerations extend to the EV's battery system, often located beneath the vehicle floor. Battery protection measures, such as robust enclosures and crash barriers, are implemented to prevent damage or thermal runaway in the event of a collision.

EVs are equipped with various safety features, including airbags, seat belts, and reinforced cabin structures, to protect occupants from impact forces and mitigate injury risks. EV manufacturers conduct comprehensive crash tests and simulations to evaluate vehicle crashworthiness according to industry standards and regulatory requirements. These tests assess frontal, side, and rear impacts, as well as rollover protection to ensure overall vehicle safety [15].

Crashworthiness is fundamental to EV safety and significantly influences consumer perceptions and adoption rates. As EVs continue to gain popularity, manufacturers are continually innovating and investing in crash-worthy designs and safety technologies to instill confidence among consumers regarding the safety of EVs. Additionally, regulatory agencies and safety organizations are vital in establishing and enforcing safety standards to ensure that EVs meet stringent crashworthiness requirements and provide high occupant protection.

EV Inlet

An EV inlet is a physical interface on an EV or HEV that connects to an external charging station via a charging cable. It serves as the receptacle for the charger plug, facilitating the transfer of electrical power from the grid to the vehicle's battery system.

The EV inlet is a critical component in the charging infrastructure of EVs and HVs, designed to meet various international standards that ensure compatibility, safety, and efficiency (Figure E.4). It is specifically engineered to handle different power levels and types of charging, including AC and DC charging [7].

FIGURE E.4 There are many connector types for vehicle charging.

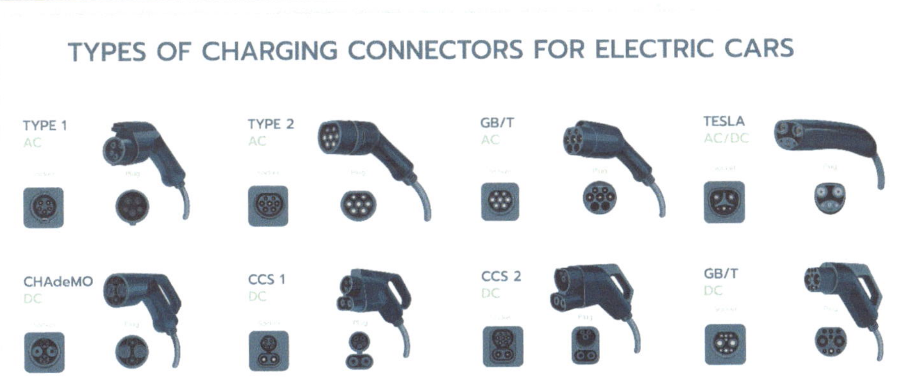

Liudmyla Militsyna/Shutterstock.com.

EV Max Currents

EV max currents refers to the maximum electrical current that EVs and HVs can draw from a charging station during the charging process. This specification ensures the vehicle's battery system's safe, efficient, and rapid charging.

The max currents that an EV or HEV can handle are determined by both the vehicle's OBC system and the specifications of the charging equipment used. These currents are typically measured in amperes (A) and are a key factor in determining the charging speed and overall time required to charge the vehicle's battery.

EV Max Voltage L1N

EV max voltage L1N refers to the maximum voltage level that an EV or HEV can accept from a single-phase connection between line 1 (L1) and neutral (N) during charging. This specification is crucial for determining the compatibility of EV charging equipment with residential or commercial power supplies [7].

The max voltage L1N is a key parameter for EV chargers, particularly in standard single-phase electrical supply settings. It indicates the upper voltage limit that can safely be handled by the vehicle's OBC from a single-phase source. Understanding this limit is essential for ensuring the safety and efficiency of the charging process.

EV Min Currents

EV min currents refers to the minimum current levels that EVs and HEVs can draw from a charging station during the charging process [14]. This specification ensures compatibility between the vehicle's charging system and the charging station's electrical infrastructure [7].

The minimum current level is a key parameter in designing and operating charging systems for EVs and HEVs. It ensures that the charging process remains efficient and safe under various conditions. This parameter is particularly important in scenarios where multiple vehicles are charging simultaneously or when the power supply is limited.

EV MVLL (Maximum Voltage Low Limit)

The EV MVLL refers to the lowest permissible voltage threshold that an EV or HEV battery can safely operate at during discharge. This limit is crucial for maintaining battery health and ensuring the vehicle's safety and efficiency [7].

EV MVLL is a parameter set by manufacturers to prevent the battery from discharging below a certain voltage level, which can lead to irreversible damage, reduced battery life, and potential safety risks such as battery failure or thermal events. The limit is typically specified in volts (V) and varies depending on the battery technology, design, and the vehicle's specific requirements.

EVCC (Electric Vehicle Communication Controller)

The EVCC is a crucial component of EVs. It facilitates communication between the vehicle's onboard systems and external charging infrastructure or intelligent grid networks [16]. As an interface, the EVCC enables the EV to communicate with charging stations, grid operators, and other vehicles to exchange information related to charging status, power demand, energy flow, and V2G functionalities [17]. The EVCC typically incorporates communication protocols such as ISO 15118, CCS, or Open Charge Point Protocol (OCPP) to enable seamless interoperability and compatibility with diverse charging infrastructure [18]. Additionally, the EVCC may also manage authentication, billing, and cybersecurity functions to ensure the secure and efficient operation of the EV within the broader electric mobility ecosystem [12].

EVHS (Electric Vehicle Home Charge Scheme)

The EVHS is a government incentive program that promotes the adoption of EVs by providing financial support for installing home charging stations. The scheme, typically administered by governmental agencies or relevant authorities, offers grants or subsidies to EV owners to help cover the costs associated with purchasing and installing a dedicated charging point at their residential property [7].

The EVHS encourages EV adoption by making it more convenient and accessible for individuals to charge their vehicles at home. Thus, it reduces reliance on public charging infrastructure and overcomes barriers to EV ownership, such as range anxiety. By enabling EV owners to recharge their vehicles overnight or during low electricity demand, home charging stations contribute to electric transportation's overall sustainability and convenience [14].

EV/PHEV Charging System

An EV or PHEV charging system refers to the infrastructure and components used to recharge the battery of an EV or PHEV. This system encompasses various charging methods, connectors, and equipment designed to power the vehicle's battery, allowing for efficient and convenient charging [3, 19].

The components of the system are as follows:

- Power generation
- Power distribution (power lines)
- EVSE
- Connector and vehicle plug
- Onboard interface electronics

For more information, see **J1715** [9].

EV Plugs

See **Type 1 Plug** and **Type 2 Plug**.

EV Receptacles

See **Type 1 Plug** and **Type 2 Plug**.

EV Requested Currents

EV requested currents refers to the specific amperage requirements that an EV or HEV communicates to a charging station during the process of recharging its battery. This parameter is crucial for ensuring that the charging process is conducted safely and efficiently, matching the vehicle's charging capability with the electrical capacity of the charging infrastructure.

The communication of EV requested currents is part of the broader protocol used in EVSE to facilitate effective interaction between the vehicle and the charging station. This ensures that the current supplied during charging does not exceed the limits of what the vehicle's OBC can handle, thereby preventing potential damage to the battery or the charging equipment.

- **Communication Standards:** EV requested currents use standardized communication protocols such as SAE J1772 or CHAdeMO to ensure compatibility and safety across different makes and models of vehicles and charging stations [7, 20].
- **Adjustable Current Requests:** Depending on the state of the battery, the EV can request variable currents to optimize charging speed and battery health. This might include requesting lower currents toward the end of the charging cycle to "top off" the battery more gently.

- **Safety Mechanisms:** The request for specific currents is integral to preventing overcurrent situations that could lead to overheating or electrical hazards [21].
- **Smart Charging Integration:** In smart charging scenarios, EV requested currents can be dynamically adjusted based on real-time energy prices, grid demand, or renewable energy availability.
- **Regulatory Compliance:** Ensuring that the charging process adheres to local electrical standards and regulations.

EV Response Error

An EV response error refers to any failure or malfunction in the communication or operational responses of EVs and HEVs to external commands or internal system triggers. This can include errors in processing charging instructions, drivetrain malfunctions, or failures in V2G communication systems.

EV response errors can occur for various reasons, such as software glitches, hardware malfunctions, incompatible charging equipment, or communication breakdowns between the vehicle and smart infrastructure (e.g., charging stations and smart grids) [7, 22]. Such errors can affect the vehicle's ability to charge properly, respond to grid demands, or execute user commands efficiently [18, 23].

- **Charging Errors:** Problems that prevent the vehicle from initiating, completing, or properly managing the charging process.
- **Drivetrain Errors:** Malfunctions that affect the performance of the electric motor, BMSs, or other components critical to the vehicle's propulsion.
- **Communication Errors:** Failures in the data exchange between the vehicle and external systems like charging stations or smart grid networks, often due to protocol mismatches or signal interference.
- **Software and Firmware Bugs:** Glitches in the vehicle's onboard software or firmware updates that lead to unexpected behaviors or system crashes.
- **Sensor and Instrumentation Failures:** Issues with sensors or electronic instruments that provide inaccurate data or fail to monitor system health effectively.

EVSE (Electric Vehicle Service Equipment)

EVSE refers to the hardware systems and related infrastructure required to safely connect an EV to a power source for charging (Tables E.2 and E.3). EVSE is a critical component of the EV ecosystem, serving as the intermediary that ensures the transfer of electrical energy from an electricity source to the EV's battery is controlled, safe, and efficient. It encompasses various charging stations, from simple residential wall-mounted units to advanced commercial and public charging stations offering fast charging capabilities.

TABLE E.2 EVSE definition of vehicle states [7].

State designation	V EVSE (VDC nominal)[5]	V vehicle (VDC nominal)[5]	Description of vehicle/EVSE state
State A	12.0[1]	0[1]	Vehicle not connected.
State B1	9.0[1]	9.0[1]	Vehicle connected/not ready to accept energy. EVSE not ready to supply energy.
State B2	9.0[2,3]	9.0[2,3]	Vehicle connected/not ready to accept energy. EVSE capable to supply energy.
Stale C	6.0[2]	6.0[2]	Vehicle connected/ready to accept energy/ indoor charging area ventilation not required. EVSE capable to supply energy.
State D	3.0[2]	3.0[2]	Vehicle connected/ready to accept energy/ indoor charging area ventilation required. EVSE capable to supply energy.
State E[4]	0	0	EVSE disconnected from vehicle/EVSE disconnected from utility, EVSE loss of utility power or control pilot short to control pilot reference.
State F	−12.0[1,6]	−12.0[1,6]	Other EVSE problem.

[1] Static voltage.
[2] Positive portion of 1 kHz square wave, measured after transition has fully settled.
[3] The transition from State B1 to State B2 begins as a static DC voltage which transitions to PWM upon the EVSE detection of vehicle connected/not ready to accept energy and EVSE capable to provide energy.
[4] EVSE is not required to actively generate State E.
[5] Voltage measured by EVSE/Vehicle as shown in Figure 1.
[6] Optional state. The EVSE may enter State F upon detecting a self-diagnosed fault that prevents the EVSE from delivering power. Once the EVSE is in STATE F, self-restoring (automatic restart) is not required. This option would require user intervention to reset the EVSE to restore normal operation. If the EVSE decides to self-restore while connected to an EV/PHEV, a maximum of 20 retries shall be allowed with a 15-min minimum time interval between retries. If the EVSE does not recover within a maximum of 20 retries, user intervention is required to reset the EVSE to restore normal operation. EVSE equipment should make every effort to recover from state F when possible.

TABLE E.3 A list of the parameters associated with the EVSE [24].

61850 Term	IEEE 1547 Parameter
WMaxEVSE	Active power setting in watts at unity power factor
ChaWMaxEVSE	Maximum active power charge setting in watts
IvarMaxEVSE	Maximum injected reactive power setting in vars
AvarMaxEVSE	Maximum absorbed reactive power setting in vars
VRefEVSE	Nominal AC voltage setting in rms volts
VRefOfsEVSE	VRef offset in rms volts (note special SAE J3072 use)
VMaxEVSE	Maximum AC voltage setting in rms volts
VMinEVSE	Minimum AC voltage setting in rms volts
UpdateTimeEVSE	UTC time stamp

EVSE systems typically include a power cable, a wall-mounted or standalone unit housing the electrical circuitry, and a connector that plugs into the EV. These systems are designed to manage the charging process, including communication with the vehicle to ensure compatibility, control of electrical current to prevent overcharging, and monitoring the charging session for safety and efficiency.

Advanced EVSE can also support bidirectional charging, enabling vehicles to return energy to the grid or home energy systems, a feature known as V2G or V2H technology (Figure E.7).

FIGURE E.7 There are multiple vehicle charging approaches [25].

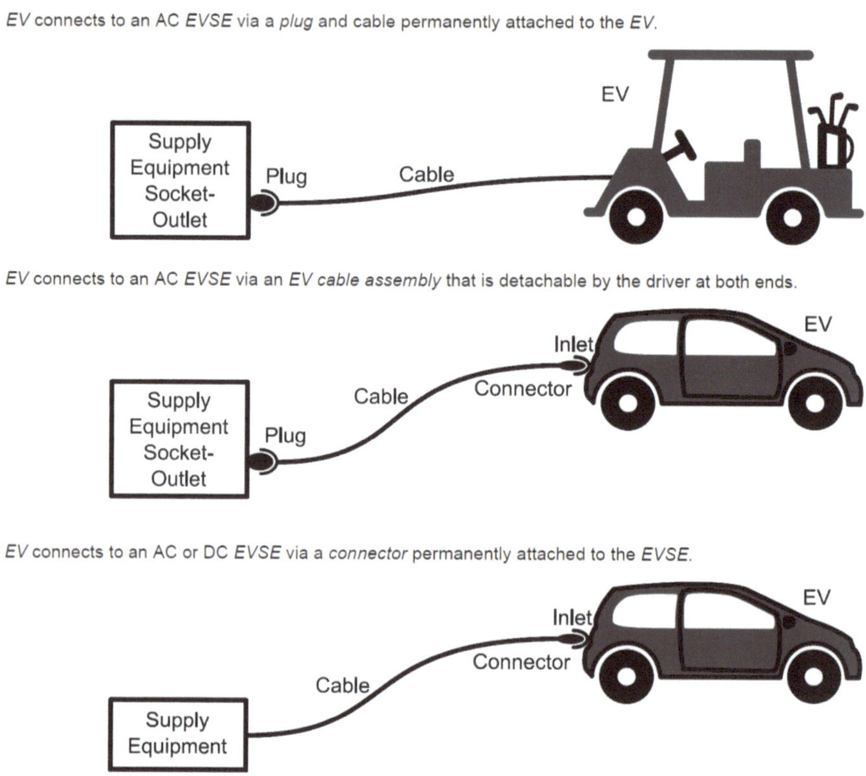

EVSE (Electric Vehicle Service Equipment) Levels

EVSE levels are categorized into different levels based on the voltage and current they provide, which affects the speed at which a vehicle's battery is charged. Understanding these levels is crucial for both consumers and industry professionals.

Level 1 Charging:

- **Voltage:** 120 V (standard household outlet).
- **Current:** Typically, up to 15 amps.
- **Power:** Up to 1.8 kW.
- **Charge Time:** Provides about 2 to 5 mi of range per hour of charging, making it suitable for overnight charging or topping off an EV's battery at home.

Level 2 Charging:

- **Voltage:** 208 to 240 V.
- **Current:** Up to 80 amps.
- **Power:** Typically, between 3.3 and 19.2 kW.
- **Charge Time:** Can provide about 10 to 60 mi of range per hour of charging, significantly faster than Level 1. Level 2 chargers are standard in residential settings, public parking areas, and workplaces.

DC Fast Charging or Level 3 Charging:

- **Voltage:** Typically, 480 V or higher.
- **Current:** Typically, up to 400 amps.
- **Power:** 50 to 350 kW, with some newer installations exceeding this range.
- **Charge Time:** Can provide 60 to 100 mi of range in about 20 min of charging. These chargers are typically found along highways and are used for rapid charging during long trips.

EVSP (Electric Vehicle Service Provider)

An EVSP is a company or organization that offers a range of services tailored specifically to the needs of EV owners and operators. EVSPs support EV adoption and widespread use by providing comprehensive charging infrastructure, maintenance services, and customer support.

EVSPs deploy and manage charging stations across various locations, including public spaces, workplaces, residential complexes, and commercial properties. These charging stations may offer different speeds and access methods, such as Level 2 AC charging or fast charging.

EVSPs often operate a network of charging stations, enabling EV drivers to conveniently locate, access, and pay for charging services through mobile apps, RFID cards, or subscription services.

EVSPs offer maintenance services for charging infrastructure, ensuring that charging stations are operational and well-maintained. They also provide customer support to assist EV drivers with charging-related queries, technical issues, and billing inquiries.

Some EVSPs offer additional services such as energy management solutions, renewable energy integration, V2G capabilities, and fleet management tools to enhance the overall experience for EV owners and operators.

EV Status Initialization

EV status initialization refers to the process through which an EV or HEV establishes its readiness state by performing self-checks and initializing system components upon start-up. This process ensures that all vehicle systems are operational and ready to function safely and efficiently before driving commences.

The initialization process is crucial for the safety and functionality of EVs and HVs. It involves diagnostic checks assessing the vehicle's battery status, power systems, software integrity, and connectivity with external devices. This routine is part of the broader vehicle management system that monitors and controls various aspects of vehicle operation, from battery management to software updates.

- **System Self-Checks:** The vehicle performs automatic diagnostics on critical systems, including the drive train, BMS, and safety systems, to ensure they function correctly.
- **Battery Status Check:** The battery's status, such as charge level, health, and temperature, is verified to ensure optimal performance and safety.
- **Software Validation:** The vehicle checks for software integrity and updates, ensuring all onboard systems run the latest and most secure software versions.
- **Sensor Calibration:** Sensors involved in driving assistance systems and vehicle operations are calibrated to ensure accurate readings and functionality.
- **Communication Systems Check:** It verifies the functionality of communication systems, including V2G and V2X communications, crucial for integrating the vehicle into the smart grid and intelligent transportation systems.

EV Status OP (Operating System)

EV Status OP is a sophisticated software system integrated into EVs to monitor and display real-time information about the vehicle's operating status. This system provides drivers with essential data regarding the EV's performance, battery health, charging status, and other vital parameters, enhancing the overall driving experience and facilitating efficient vehicle management.

EV Status OP continuously monitors various aspects of the vehicle's operation, including battery charge level, range estimation, energy consumption, and system diagnostics. The system presents information to the driver through an intuitive and user-friendly interface, typically displayed on the vehicle's dashboard or infotainment screen. EV Status OP provides insights into the health and condition of the EV's battery pack, including SOC, temperature, and overall battery performance over time.

Drivers can conveniently check the status of ongoing charging sessions, including charging rate, time remaining until full charge, and any charging errors or warnings. The system utilizes advanced algorithms to calculate and display an accurate estimation of the remaining driving range based on the current battery charge level, driving behavior, and environmental factors. EV Status OP conducts self-diagnostic checks and alerts drivers to any potential issues or maintenance requirements, helping to prevent unexpected breakdowns and ensuring optimal vehicle performance.

EVT (Electric Variable Transmission)

EVT is a type of transmission system used primarily in HEVs and some EVs that combines traditional mechanical gears with electric motor generators to provide a seamless range of gear ratios [26]. This system allows optimal power and efficiency by dynamically adjusting the power flow between the engine, electric motors, and wheels. An EV uses electric motors as the power source. The motors are coupled to mechanical drive train with the help of transmission system [27].

EVT is designed to optimize fuel efficiency and enhance vehicle performance by leveraging the combustion engine and electric motors. The system can vary the gear ratio continuously, unlike conventional transmissions that have a fixed number of gear ratios. This capability not only improves the driving experience by providing smoother acceleration and deceleration but also helps in reducing emissions and improving fuel economy.

Key Features:

1. **Continuously Variable Transmission (CVT) Capability:** EVT operates similarly to a CVT but with the addition of electric motors to assist or take over from the combustion engine.

2. **Regenerative Braking:** It utilizes the electric motor as a generator during braking or coasting to convert kinetic energy into electrical energy, which is then stored in the battery.
3. **Reduced Mechanical Complexity:** By integrating electric motors, EVTs can reduce the complexity and number of mechanical components compared to traditional transmissions.
4. **Dual-Mode Operation:** Some EVTs can operate in dual modes, switching between purely electric propulsion and combined engine-motor propulsion, depending on driving conditions and battery charge levels.

Types of EVTs:

1. **Single-Mode EVT:** It utilizes a single electric motor generator integrated into a simpler transmission mechanism. This type is typically used in mild-HVs and some full-HVs models where less variability in the transmission range is required.
2. **Dual-Mode EVT:** It is more complex and provides two distinct modes of operation, allowing the vehicle to switch between modes based on driving conditions. This type is often found in full-HVs and PHEVs, offering a wider range of power and efficiency benefits.
3. **Power-Split EVT:** Commonly used in many full-HVs, this type uses two motor generators. It effectively splits power between the engine and electric motors, allowing for an efficient balance of power distribution, enhancing both performance and fuel economy.

REFERENCES

[1]. S. Cash, Q. Zhou, H. Xu, O. Olatunbosun, R. Shaw and S. Davis, "New Traction Motor Sizing Strategy for an HEV/EV Based on an Overcurrent-Tolerant Prediction Model," *IET Intelligent Transport Systems*, 13, 168, 2019.

[2]. T. Saito and K. Miyamoto, "*Vibrational Analysis Method on High-Frequency Electric-Drive Motor Noise*," SAE Publishing, Warrendale, 2020.

[3]. Society of Automotive Engineers, "J1766 Recommended Practice for Electric, Fuel Cell and Hybrid Electric Vehicle Crash Integrity Testing," SAE Publishing, Warrendale, 2014.

[4]. ISO, "ISO 6469-1: Electrically Propelled Road Vehicles—Safety Specifications—Part 1: On-Board Rechargeable Energy Storage System (RESS)," ISO Publishing, Geneva, 2019.

[5]. International Electrotechnical Commission (IEC), "Electric Vehicle Conductive Charging System - Part 23: DC Electric Vehicle Charging Station (IEC 61851-23:2023)," International Electrotechnical Commission Publishing, Geneva, 2023.

[6]. Society of Automotive Engineers, "J3042 Measuring Properties of Li-Ion Battery Electrolyte," SAE Publishing, Warrendale, 2021.

[7]. Society of Automotive Engineers, "J1772 SAE Electric Vehicle and Plug-In Hybrid Electric Vehicle Conductive Charge Coupler," SAE Publishing, Warrendale, 2024.

[8]. S. Davis, *Managing Electric Vehicle Power*, Warrendale: SAE Publishing, 2020.

[9]. Society of Automotive Engineers, "J1715 Hybrid Electric Vehicle (HEV) and Electric Vehicle (EV) Terminology," SAE Publishing, Warrendale, 2022.

[10]. Society of Automotive Engineers, "J551 Vehicle Electromagnetic Immunity - Electrostatic Discharge (ESD)," SAE Publishing, Warrendale, 2020.

[11]. Society of Automotive Engineers, "J1113-13 Electromagnetic Compatibility Measurement Procedure for Vehicle Components - Part 13: Immunity to Electrostatic Discharge," SAE Publishing, Warrendale, 2015.

[12]. Society of Automotive Engineers, "J2293 Energy Transfer System for Electric Vehicles - Part 2: Communication Requirements and Network Architecture," SAE Publishing, Warrendale, 2014.

[13]. Society of Automotive Engineers, "J3073 Battery Thermal Management," SAE Publishing, Warrendale, 2016.

[14]. Society of Automotive Engineers, "J3068 Electric Vehicle Power Transfer System Using a Three-Phase Capable Coupler," SAE Publishing, Warrendale, 2024.

[15]. Society of Automotive Engineers, "J3121 Hydrogen Vehicle Crash Test Lab Safety Guidelines," SAE Publishing, Warrendale, 2022.

[16]. Society of Automotive Engineers, "J2931-4 Broadband PLC Communication for Plug-In Electric Vehicles," SAE Publishing, Warrendale, 2023.

[17]. Society of Automotive Engineers, "J2836 Use Cases for Diagnostic Communication for Plug-In Electric Vehicles," SAE Publishing, Warrendale, 2021.

[18]. Society of Automotive Engineers, "J2847-6 Communication for Wireless Power Transfer Between Light-Duty Plug-In Electric Vehicles and Wireless EV Charging Stations," SAE Publishing, Warrendale, 2020.

[19]. Society of Automotive Engineers, "J2953-4 Plug-In Electrical Vehicle Charge Rate Reporting and Test Procedures," SAE Publishing, Warrendale, 2024.

[20]. CHAdeMO EU Secretariat, "CHAdeMO," April 2019, accessed June 28, 2024, https://www.chademo.com.

[21]. Society of Automotive Engineers, "J2990 Hybrid and Electric Vehicle Safety Systems Information Report," SAE Publishing, Warrendale, 2020.

[22]. Society of Automotive Engineers, "J3068 Identification of Vehicles and Supply Equipment for Conductive AC Charging," SAE Publishing, Warrendale, 2023.

[23]. Society of Automotive Engineers, "J2847-2 Communication between Plug-In Vehicles and Off-Board DC Chargers," SAE Publishing, Warrendale, 2023.

[24]. Society of Automotive Engineers, "J3072 Interconnection Requirements for Onboard, Grid Support Inverter Systems," SAE Publishing, Warrendale, 2021.

[25]. Society of Automotive Engineers, "J3400 NACS Electric Vehicle Coupler," SAE Publishing, Warrendale, 2023.

[26]. J. F. Shore, A. I. Christodoulias, A. S. Kolekar, F. E. Lockwood and A. Kadiric, *Prediction of Electric Vehicle Transmission Efficiency Using a New Thermally Coupled Lubrication Model*, Warrendale: SAE Publishing, 2022.

[27]. S. Sirohi, S. Yadav, B. Ashok, V. R. Babu, C. Kavitha and K. G. Nantha, *Structural Analysis of Electric Vehicle Transmission - Mounts and Casing for Different Materials*, Warrendale: SAE Publishing, 2017.

F

Fast Charging

Fast charging refers to a method that significantly reduces the time required to replenish the battery's energy compared to standard charging methods. Fast charging technologies are designed to accommodate the rapid intake of electrical current, allowing EV batteries to charge from 20 to 80% capacity in as little as 20 to 30 min, depending on the battery and charger specifications.

High-power charging stations deliver a much higher voltage and current than typical home charging stations. These stations, often found along highways and in public areas, use DC rather than the AC used in conventional slower charging setups.

Several standards for fast charging exist, including CCS, CHAdeMO, and Tesla's Supercharger. Each is compatible with different vehicle makes and models. An EV's compatibility with a particular type of fast charger depends on its onboard charging hardware and software capabilities.

While fast charging is convenient for long-distance travel, frequent use can accelerate Li-ion batteries' degradation. Vehicle manufacturers and battery engineers continuously work to improve battery life and fast charging capabilities to mitigate such effects.

FCHEV (Fuel Cell Hybrid Electric Vehicle)

An FCHEV is an advanced type of EV that utilizes hydrogen fuel cells and a high-voltage battery to power an electric motor (Figure F.1). Unlike conventional HVs, which typically rely on ICEs and batteries, FCHEVs employ fuel cells as the primary source of electricity generation, with the battery serving as a secondary ESS.

FIGURE F.1 An example of the systems of a hydrogen fuel cell vehicle (FCV).

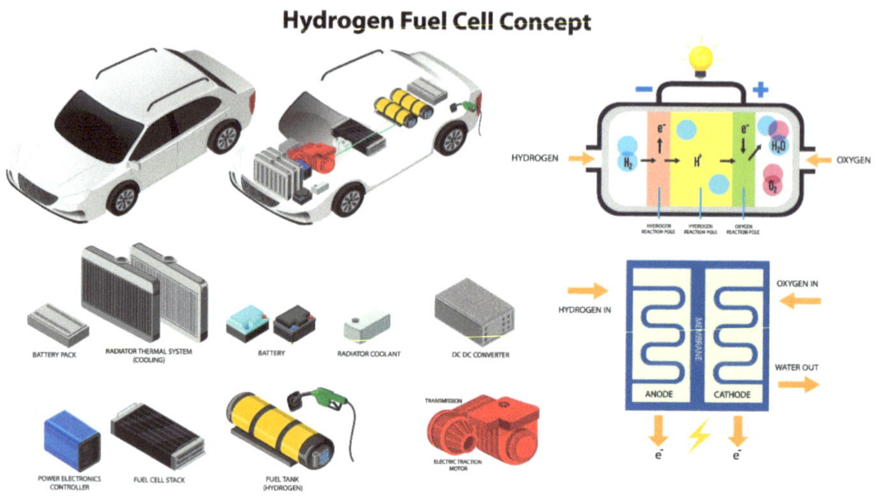

BlueRingMedia/Shutterstock.com.

An FCHEV is an innovative type of EV that utilizes fuel cell technology to generate electricity on board, which powers an electric motor to drive the vehicle. Unlike traditional BEVs that rely solely on rechargeable batteries, FCHEVs produce electricity through an electrochemical reaction between hydrogen and oxygen, with water vapor as the only by-product.

Flywheel Assist System

A flywheel assist system in EVs refers to a mechanical energy storage device that stores rotational energy in a rapidly spinning wheel. This system is designed to enhance the efficiency and performance of EVs by capturing kinetic energy during braking and releasing it to aid acceleration [1].

In EVs, the flywheel assist system operates on the principle of regenerative braking (Figure F.2). When the vehicle decelerates, the kinetic energy, which would otherwise be lost as heat through the brakes, is transferred to the flywheel, causing it to spin at high speeds. Upon acceleration, the stored energy in the flywheel is converted back to mechanical energy, boosting the vehicle's motor and reducing the demand for the battery. This process not only conserves energy but also extends the vehicle's range [2].

FIGURE F.2 Architecture that includes flywheel [2].

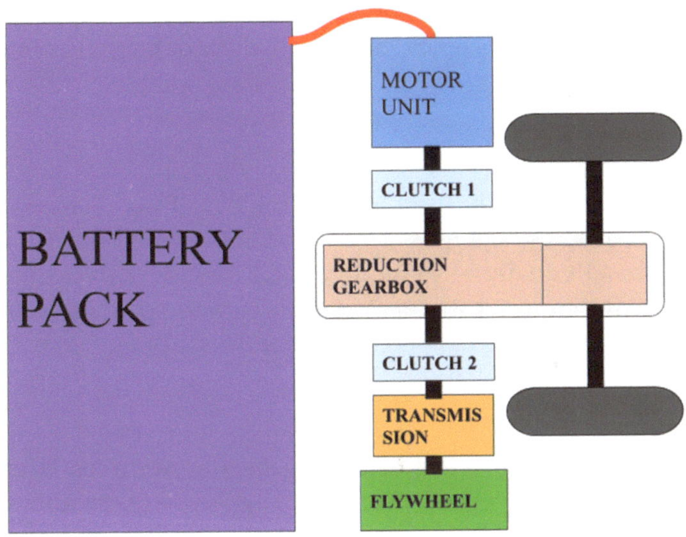

Four-Quadrant Inverter

A four-quadrant inverter is an advanced electrical device used in EVs to convert DC from the battery into AC to drive the electric motors. It can operate in all four quadrants of the power flow, allowing for both motoring and regenerative braking operations.

The term "four-quadrant" refers to the inverter's ability to handle power in four ways: forward motoring, forward braking (regenerative), reverse motoring, and reverse braking. This functionality is crucial for EVs as it provides the necessary AC power to the motors for propulsion and allows the vehicle to recuperate energy during braking, enhancing energy efficiency, and extending the range.

A typical four-quadrant inverter includes components like IGBTs or metal-oxide-semiconductor field-effect transistors (MOSFETs), which are used for switching the current flow at high frequencies. The inverter also integrates advanced control algorithms typically managed by an onboard microcontroller or digital signal processor to precisely control the timing and duration of the current flow to the motors.

The four-quadrant inverter plays a pivotal role in the drivetrain system in EVs. It drives the electric motor during vehicle acceleration and manages the energy recaptured during deceleration. This dual capability supports the vehicle's performance and efficiency, making it a fundamental component in modern EV architecture.

Frunk/Froot

A "frunk" (a portmanteau of "front" and "trunk") or "froot" (a blend of "front" and "boot") refers to the storage space located in the front of an EV where the engine would traditionally be found in an ICE vehicle. This feature is made possible due to the compact size of electric motors compared to conventional engines, allowing manufacturers to repurpose the space for cargo.

Fuel Cell

A fuel cell is an electrochemical device that generates electricity by converting the chemical energy of a fuel and an oxidizing agent directly, without combustion (Figure F.3). In EVs, fuel cells serve as an alternative power source to conventional batteries, providing continuous electrical power for propulsion.

FIGURE F.3 An example of the inner workings of a hydrogen fuel cell.

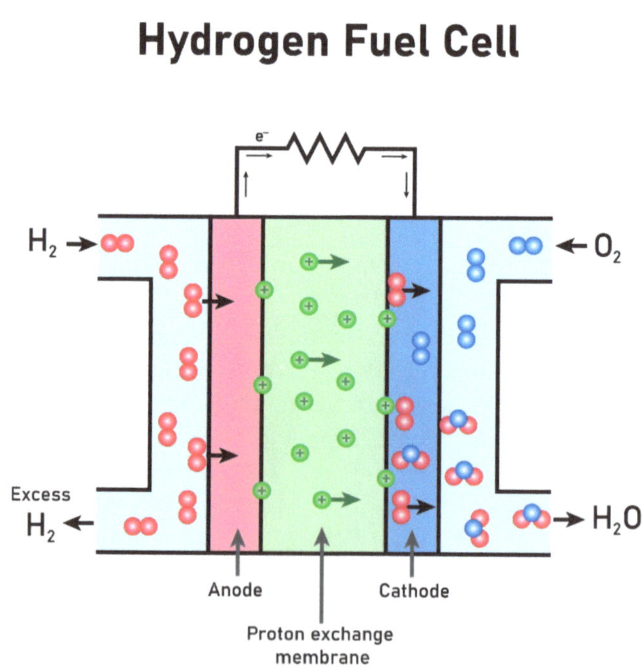

Dream01/Shutterstock.com.

Types of Fuel Cells Used in EVs:

1. **Proton Exchange Membrane Fuel Cell (PEMFC):**
    - **Description:** The most common type of fuel cell used in vehicles, PEMFC operates at relatively low temperatures (around 80°C), making it suitable for quick start-up times and varying power outputs required for automotive applications.
    - **Advantages:** High power density and lower operating temperatures.
    - **Applications:** Predominantly used in passenger cars and light commercial vehicles.
2. **Solid Oxide Fuel Cell (SOFC):**
    - **Description:** This type of fuel cell operates at high temperatures (between 600 and 1000°C). While less common in mainstream vehicles due to their high operating temperature and slower start-up times, ongoing research aims to adapt SOFCs for vehicle use due to their high efficiency and fuel flexibility.
    - **Advantages:** Can use multiple types of fuels and high electrical efficiency.
    - **Applications:** Potential use in heavy-duty vehicles and stationery charging applications for EVs.
3. **Alkaline Fuel Cell (AFC):**
    - **Description:** Once prominent in space applications, AFCs are being explored for automotive use. They operate on compressed hydrogen and oxygen, offering a clean by-product—water. However, their sensitivity to CO_2 requires clean operating environments.
    - **Advantages:** High efficiency and low operating temperatures.
    - **Applications:** Experimental and niche automotive applications.

Fusible Link

A fusible link is a safety device used in EVs and other electrical systems to protect against overcurrent conditions and prevent electrical fires or damage to components. It consists of a specially designed wire or strip engineered to melt and break the circuit in case of excessive current flow, effectively interrupting the power supply and preventing further damage.

The primary function of a fusible link is to safeguard electrical circuits and components from damage caused by overloads or short circuits. When the current passing through the link exceeds its rated capacity due to a fault or malfunction, the fusible link heats up rapidly, causing the fusible material to melt and create an open circuit. This action disconnects the affected circuit, preventing the flow of excessive current and mitigating the risk of electrical hazards or component failure.

Fusible links are essential safety components in EVs. They protect sensitive electronic systems, batteries, wiring harnesses, and other critical components from thermal damage or catastrophic failure. Their ability to quickly interrupt current flow under abnormal conditions helps maintain the integrity and reliability of the EV's electrical infrastructure, enhancing overall safety and performance.

Fuses

Fuses are vital components in designing and operating EVs and HVs. They protect against electrical hazards and play a crucial role in the vehicle's overall reliability and performance. Proper fuse selection and placement ensure that all parts of the vehicle's electrical system function within their designed parameters, mitigating risks associated with electrical failures.

In EVs and HVs, fuses are integrated into various parts of the electrical system, including the battery pack, power inverter, and onboard charging system. The fuses are specifically engineered to handle the high voltages and currents typical in these vehicles, which can range up to several hundred volts for the main traction battery.

The primary function of fuses in EVs and HVs is to safeguard the electrical integrity of the vehicle by interrupting the power supply when the current flow exceeds safe levels. This interruption helps prevent potential hazards caused by electrical faults, thereby ensuring the vehicle's occupants' safety and the electrical components' longevity.

Types:

1. **Blade Fuses:** Compact and commonly used for low to moderate current applications.
2. **Cartridge Fuses:** Used for higher current circuits, often found in main battery connections.
3. **Semiconductor Fuses:** Designed to protect high-speed electronic components and can break the circuit in microseconds.

REFERENCES

[1]. Society of Automotive Engineers, "J1715 Hybrid Electric Vehicle (HEV) and Electric Vehicle (EV) Terminology," SAE Publishing, Warrendale, 2022.

[2]. P. S. Venkatesh, V. Chandran and S. Anil, *Study of Flywheel Energy Storage in a Pure EV Powertrain in a Parallel Hybrid Setup and Development of a Novel Flywheel Design for Regeneration Efficiency Improvement*, Warrendale: SAE Publishing, 2021.

G

G2V (Grid-to-Vehicle)

G2V integration refers to the bidirectional communication and energy transfer capabilities between EVs or HEVs and the electrical grid infrastructure. This concept enables EVs to consume electricity from the grid for charging and provide stored energy back to the grid when needed, effectively transforming EVs into mobile energy storage units.

The primary function of G2V integration is to allow EVs to charge their batteries from the electrical grid. This involves charging stations or connectors that directly interface the vehicle's onboard charging system and the grid's power supply.

G2V systems often incorporate smart charging and energy management algorithms. These algorithms optimize charging schedules based on factors such as electricity prices, grid demand, renewable energy availability, and user preferences, maximizing efficiency and cost-effectiveness.

G2V integration contributes to grid stability by offering DR capabilities. EVs can adjust their charging or discharging patterns in response to grid signals, helping to balance supply and demand and mitigate grid fluctuations. G2V systems facilitate the integration of renewable energy sources into the grid. EVs can charge when renewable energy generation is high, reducing reliance on fossil fuels and promoting sustainable energy practices. In emergencies or grid outages, EVs with G2V capabilities can be backup power sources for homes, businesses, or critical infrastructure. This enhances energy resilience and provides a valuable resource during unforeseen disruptions.

See also **V2G (Vehicle-to-Grid)**.

GA (Ground Assembly) (WPT)

In EVs, a GA is a crucial component of WPT systems.

See **Inductive Charging System**.

General FCV (Fuel Cell Vehicle) Safety

FCVs are EVs that use hydrogen fuel cells to generate electricity, powering an electric motor to propel the vehicle. While FCVs offer numerous environmental benefits, such as zero tailpipe emissions and increased energy efficiency, safety considerations are paramount. The electrical considerations are similar to those of EVs, coupled with the dangers of hydrogen as a fuel.

GHG (Greenhouse Gas)

GHGs are gases in the Earth's atmosphere that contribute to the greenhouse effect by trapping heat radiated from the Earth's surface (Figure G.1). They play a significant role in climate change by enhancing the planet's natural greenhouse effect, leading to increased global temperatures. In the context of EVs, reducing GHG emissions is a critical environmental benefit compared to traditional ICE vehicles.

FIGURE G.1 An illustration of the sources of greenhouse gases.

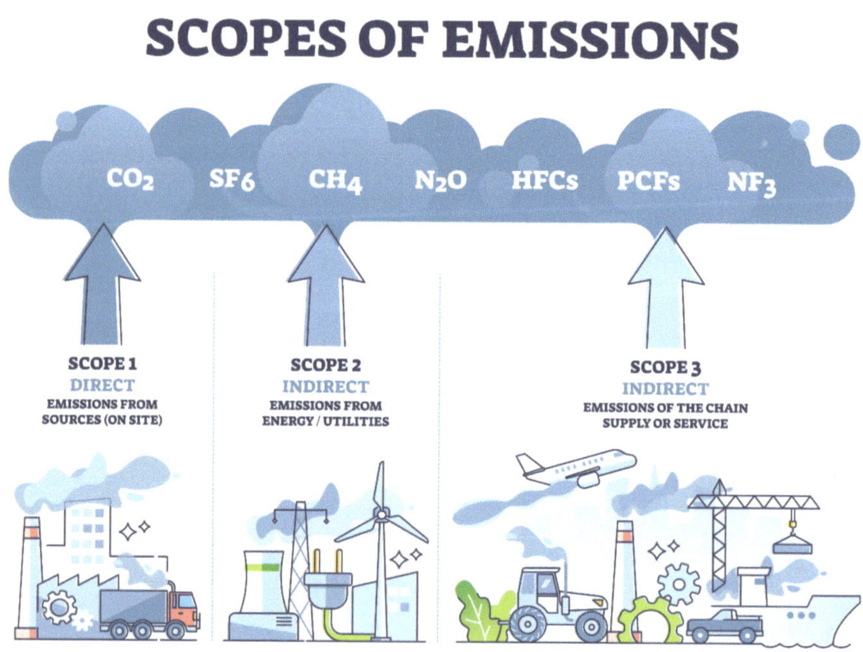

VectorMine/Shutterstock.com.

Reducing GHG emissions is essential for mitigating climate change and transitioning to a more sustainable energy future. EVs, with their cleaner and more efficient mode of transportation, play a crucial and leading role in this transition. They significantly contribute to lower GHG emissions and environmental impact, making them a key solution in the fight against climate change.

Grid Topologies

In the context of EVs, grid topologies refer to the various configurations and architectures of the electrical grids that support charging infrastructure (Figure G.2). These topologies are crucial in determining EV charging networks' efficiency, reliability, and scalability.

FIGURE G.2 An illustration of centralized and decentralized EV charging systems.

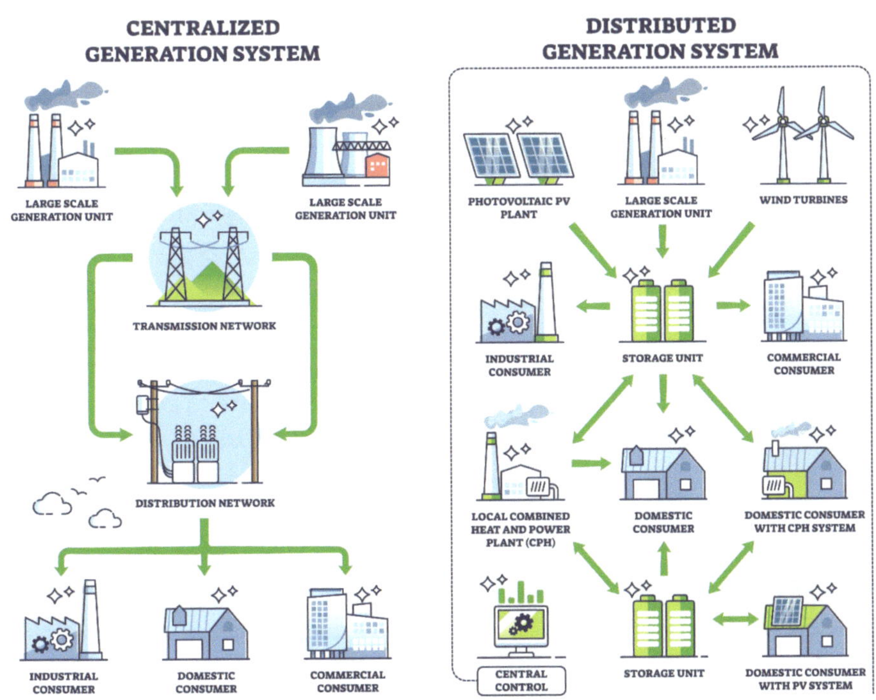

Types of Grid Topologies:

- **Centralized Grid:**
  In a centralized grid topology, power generation and distribution are managed from a central point, such as a power plant or substation. Charging stations are connected to the central grid, which distributes power to EVs based on demand and available capacity. This topology is common in traditional power distribution systems and may require upgrades to support the increasing demand for EV charging.
- **Decentralized Grid:**
  Decentralized grid topologies, which distribute power generation and storage across multiple smaller units, such as solar panels, wind turbines, and local ESSs, offer increased resilience and flexibility. However, they may require intelligent grid technologies to manage intermittent and variability in power generation, presenting both challenges and benefits.
- **Smart Grid:**

  Innovative grid topologies integrate advanced communication, control, and monitoring technologies into the electrical grid infrastructure. These grids enable bidirectional communication between EVs, charging stations, and the grid, allowing for dynamic load management, DR, and optimization of charging schedules. Smart grids facilitate the integration of renewable energy sources, grid-scale energy storage, and electric V2G capabilities, enhancing grid stability, efficiency, and sustainability.

Gross Battery Capacity

In the context of EVs, gross battery capacity refers to the total energy storage capacity of the vehicle's battery pack. It represents the maximum amount of electrical energy the battery can store and subsequently deliver to power the vehicle's electric motor(s). Gross battery capacity is typically expressed in kWh and is a fundamental metric for assessing an EV's driving range and performance capabilities.

Ground

In EVs, "ground" refers to the electrical connection between the vehicle's components and the earth, typically achieved through a conductive pathway. The ground connection serves several vital purposes in an EV, including safety, electrical system integrity, and noise reduction.

Grounding Conductor

A grounding conductor, also known as an earth conductor or ground wire, is an essential component of the electrical system in EVs. It plays a critical role in providing a safe pathway for the dissipation of electrical faults and ensuring the protection of vehicle occupants and sensitive electronic components.

The grounding conductor is a wire or metal strip connected to the EV's chassis or frame. Its primary function is establishing a direct connection between electrical components and the ground or earth. In the event of a fault, such as a short circuit or electrical surge, the grounding conductor channels excess electrical current safely into the ground, preventing the buildup of dangerous voltage levels within the vehicle.

A grounding conductor is crucial for electrical safety in EVs. It helps to mitigate the risk of electric shock to occupants and reduces the likelihood of electrical fires or damage to sensitive electronic systems. Additionally, grounding conductors are mandated by electrical codes and standards to ensure compliance with safety regulations and industry best practices.

Grounding conductors are integrated into EVs' electrical wiring and grounding systems. They are typically connected to components such as the battery, powertrain, charging system, and electrical accessories. Proper installation and maintenance of grounding conductors are essential to ensure their effectiveness in providing a reliable path for fault current to dissipate safely.

H

Hazardous Voltage Faults

Hazardous voltage faults refer to potentially dangerous electrical malfunctions or abnormalities occurring within EVs, generating hazardous voltage levels. These faults pose a safety risk to vehicle occupants, service technicians, and bystanders due to the potential for electric shock, fire, or other hazardous consequences [1]. A brief description of the types of failures follows.

A short circuit occurs when an unintended low-resistance pathway is created between high-voltage components, resulting in a sudden surge of electrical current. This can lead to overheating, arcing, and the risk of electrical fires. Failures can occur when the protective insulation around electrical conductors or components degrades or becomes damaged, allowing high-voltage electricity to encounter conductive surfaces or human body parts [2]. Ground faults occur when an unintended electrical connection is established between a high-voltage circuit and the ground (earth). This can result in excessive current flow through unintended pathways, posing a risk of electric shock or equipment damage. Overvoltage occurs when the voltage in a circuit exceeds safe operating limits, potentially causing damage to electrical components, insulation breakdown, or arcing [1].

HDEV (Hybrid Diesel Electric Vehicle)

An HDEV combines a diesel engine with electric propulsion systems to achieve improved fuel efficiency, reduced emissions, and enhanced performance. This hybrid configuration optimizes power delivery and energy management, making HDEVs popular for commercial vehicles, buses, and heavy-duty applications.

Heat Pump

A heat pump is a crucial component in EVs designed to manage the thermal control of the vehicle's interior and battery system. It transfers heat from one area to another, using refrigerant cycles to provide heating or cooling as needed. In EVs, heat pumps are crucial in improving energy efficiency, extending driving range, and enhancing overall comfort for occupants [3].

The primary function of a heat pump in an EV is to regulate the temperature inside the cabin and maintain optimal operating conditions for the battery pack. During cold weather, the heat pump can extract heat from the ambient environment or waste heat from the vehicle's systems (such as the motor or power electronics) and transfer it into the cabin to warm the interior without relying solely on energy-intensive electric heating elements. Conversely, in hot weather, the heat pump can remove heat from the cabin and dissipate it outside.

Heat pumps are integrated into the HVAC system of EVs, working with subassembly components such as refrigerant compressors, evaporators, condensers, and heat exchangers. They are controlled by the vehicle's thermal management system, which monitors temperature sensors and adjusts the heat pump's operation based on internal and external conditions.

See also **Thermal Management System**.

HEV (Hybrid Electric Vehicle)

An HEV combines two or more power sources to propel it, typically an ICE and an electric motor. HEVs are designed to optimize fuel efficiency, reduce emissions, and enhance overall performance by leveraging the strengths of both conventional ICEs and electric propulsion systems.

There are three types of hybrid systems. The parallel hybrid system, the ICE, and the electric motor are mechanically connected to the vehicle's drivetrain, allowing them to work together to drive the wheels. The electric motor assists the engine during acceleration, deceleration, or low-speed driving, reducing fuel consumption and emissions.

The second type, the series hybrid, uses an ICE to generate electricity solely and then powers an electric motor that drives the wheels. The engine operates optimally to charge the battery or provide additional power when needed, while the electric motor handles propulsion, offering smoother and quieter operation.

Finally, there is the PHEV. A PHEV is a variant of HEVs with a larger battery pack capable of being charged from an external power source, such as a charging station or electrical outlet. PHEVs offer an extended electric-only driving range, allowing drivers to operate in all-electric mode for shorter trips, reducing reliance on gasoline and lowering emissions.

High-Power SSR (Solid-State Relay)

A high-power SSR is an electronic switching device that uses semiconductor components to switch high currents and voltages in various applications without moving parts, unlike electromechanical relays. High-power SSRs are extensively used in automotive applications, particularly in EVs and HVs, for managing power distribution and control within the vehicle's electrical systems. SSRs are more reliable, provide faster switching time and reduced heat generation [4].

High Voltage

Any voltage level exceeding the standard automotive 12- or 24-V systems is high voltage. In EVs, high voltage typically refers to voltages above 60 V DC or 30 V AC, used primarily in the propulsion of EVs.

High-Voltage Automotive Wiring Assembly Design

High-voltage automotive wiring assembly design refers to the specialized engineering process of designing electrical wiring systems for EVs operating at high voltage levels [1]. These systems are crucial components of EVs, facilitating power transmission from the battery to various electrical components, including motors, inverters, chargers, and auxiliary systems.

For more information, see **J2840**, **J2344**, and **J1673** [5–7].

High-Voltage Distribution System

High-Voltage Distribution System (intra-vehicle) refers to the network of electrical components and wiring that distribute high-voltage electricity throughout the vehicle. This system is responsible for delivering power from the vehicle's high-voltage battery pack to various critical components, such as the electric motor, power electronics, and auxiliary systems like the air conditioning compressor or electric heater.

The high-voltage distribution system typically operates at voltages ranging from 200 to over 800 V, depending on the vehicle design and application. It is designed to handle the substantial power requirements of electric propulsion systems while ensuring safety and efficiency. Components within this system include high-voltage cables, connectors, junction boxes, and safety mechanisms like relays and fuses that protect against electrical faults.

Safety is a paramount concern in high-voltage distribution systems, as the high energy levels involved can be dangerous if not properly managed. As such, these systems are engineered with robust insulation, protective enclosures, and fault detection mechanisms to ensure safe operation and to isolate high-voltage components in the event of an accident or malfunction.

High-Voltage Primary Cable

A high-voltage primary cable, often referred to simply as a primary cable, is an essential component in the electrical system of EVs designed to carry high-voltage electrical currents from the battery pack to various components such as the motor, inverters, and charging system. These cables are frequently shielded and engineered to withstand high levels of electrical stress and are crucial for ensuring efficient power delivery within the EV's powertrain.

The primary function of a high-voltage primary cable is to transmit high-voltage DC or AC from the battery pack or power distribution unit to the vital electrical components responsible for propulsion, energy conversion, and charging. These cables are typically insulated with materials capable of withstanding the high voltages and currents encountered in EV systems, thereby minimizing the risk of electrical arcing, overheating, or insulation breakdown.

High-voltage primary cables are constructed using specialized materials and techniques to meet the rigorous demands of EV applications. They often feature multiple layers of insulation, such as cross-linked polyethylene (XLPE), ethylene propylene rubber (EPR), or silicone rubber, to provide excellent electrical insulation and mechanical strength. The conductors inside these cables are typically made of copper or aluminum alloys optimized for high conductivity and minimal electrical losses.

For more information, see **J2840**, **J2344**, and **J1673** [5–7].

High-Voltage Systems

High-voltage systems refers to the electrical systems and components in EVs that operate at elevated voltage levels, typically above 60 V DC. These systems are crucial for powering key EV components such as traction motors, inverters, OBCs, and ESSs, including high-voltage battery packs. The key components of the vehicle's high-voltage systems are as follows:

- High-voltage battery pack
- Traction motor
- Invertor
- OBC
- DC-DC convertor

For additional information, see **J1715** [8].

High-Voltage Withstand Capability

High-voltage withstand capability refers to the ability of electrical components, systems, or devices in an EV to withstand and operate safely under high-voltage conditions without experiencing electrical breakdown or insulation failure. This capability ensures EVs' reliability, safety, and performance, particularly in high-voltage systems such as battery packs, motor controllers, inverters, and charging infrastructure. EV components' high-voltage withstand capability is tested and specified to withstand voltage levels significantly higher than their nominal operating voltages. This ensures they can endure transient voltage spikes, surges, and fluctuations commonly encountered in EV powertrains without compromising functionality or safety. Components with high-voltage withstand capability are designed with robust insulation, shielding, and protective features to prevent electrical arcing, short circuits, and insulation breakdown under high-voltage conditions.

For more information, see **J2840** and **J553** [5, 9].

HPU (Hybrid Power Unit)

An HPU refers to a specialized powertrain system used in HEVs and PHEVs. The HPU combines multiple propulsion technologies, typically an ICE and an electric motor, to provide propulsion and optimize fuel efficiency in varying driving conditions.

The HPU operates on the principle of power blending, where the ICE and electric motor work in tandem or independently, depending on driving demands and operating modes. During low-speed or light-load conditions, the electric motor may power the vehicle, drawing energy from the onboard battery pack. Conversely, during high-speed acceleration or heavy-load scenarios, the ICE may engage to provide additional power or recharge the battery through regenerative braking.

For more information, see **J1715** [8].

HVBD (Hazardous Voltage Bus Discharge)

HVBD refers to the controlled process of discharging high-voltage buses or systems in EVs to ensure the safety of technicians, maintenance personnel, and first responders during emergencies or maintenance procedures. HVBD aims to reduce the risk of electric shock or injury by safely dissipating stored electrical energy from high-voltage components [10].

HVBD procedures are activated when an EV is involved in an accident, experiences a system malfunction, or undergoes maintenance that requires deenergized high-voltage systems [11]. This typically involves activating discharge circuits or devices that create a path to safely discharge electrical energy from high-voltage capacitors, batteries, or other components. HVBD may be initiated manually by trained personnel or automatically through onboard safety systems.

HVBD plays a crucial role in ensuring the safety of individuals interacting with EVs, especially in emergency scenarios where rapid deenergization of high-voltage systems is necessary. By effectively discharging hazardous voltage buses, HVBD minimizes the risk of electric shock, arc flash incidents, and other electrical hazards, providing a safer environment for rescue operations, maintenance work, and vehicle recovery.

HVD (High-Voltage Disconnect)

An HVD is a safety feature incorporated into EVs and HEVs to ensure the safe disconnection of high-voltage electrical systems from the vehicle's powertrain and components. This disconnect mechanism is designed to isolate the high-voltage battery pack and associated circuits during maintenance, servicing, or emergencies, reducing the risk of electric shock or injury to personnel.

The primary function of a HVD is to physically break the electrical connection between the high-voltage battery pack and the rest of the vehicle's electrical systems. When activated, the HVD interrupts the flow of high-voltage current, preventing energized components and circuits from posing a hazard to technicians, first responders, or bystanders. The disconnect may be manually operated or activated automatically in response to specific conditions, such as a collision or system fault.

Manual disconnects are operated by a manual switch or lever accessible to service personnel. Manual disconnects require deliberate action to activate and deactivate, providing control over high-voltage isolation.

Automatic disconnects, also known as emergency or activated disconnects, are triggered automatically in response to predefined events, such as airbag deployment, collision detection, or system fault detection.

HVDs are integrated into the electrical architecture of EVs and HVs, typically located near the high-voltage battery pack or power distribution unit. They may be incorporated into safety systems, such as the vehicle's OBD system or the vehicle control unit (VCU), to enable remote activation or monitoring of the disconnect status. HVDs are often accompanied by visual indicators or alarms to signal when the high-voltage system is disconnected or reconnected.

HVIL (Hazardous Voltage Interlock Loop)

An HVIL is a safety feature integrated into EVs to minimize the risk of electric shock or injury during maintenance or repair activities. It is designed to ensure that high-voltage components within the EV's powertrain are correctly isolated and deenergized when necessary, thereby safeguarding technicians and maintenance personnel from exposure to hazardous voltages [6]. The HVIL operates by establishing a closed-loop circuit that continuously monitors the status of high-voltage systems within the EV. When maintenance or repair work is required, the HVIL system detects the opening of access panels or doors, signaling the need to deactivate high-voltage components. Once triggered, the HVIL system initiates a series of interlocks and safety measures to isolate the high-voltage circuitry, including activating contactors or relays to interrupt power flow and engaging safety mechanisms to prevent accidental re-energization.

Implementing an HVIL system is crucial for ensuring the safety of service technicians and maintenance personnel working on EVs. By effectively isolating high-voltage components, the HVIL minimizes the risk of electric shock or injury, providing a safe working environment for maintenance and repair operations. Additionally, compliance with HVIL safety standards and regulations is essential for EV manufacturers to uphold industry safety standards and ensure the reliability and trustworthiness of their vehicles.

HVIL systems are typically integrated into the overall safety architecture of EVs, alongside other safety features such as high-voltage interlock systems (HVIS), insulation monitoring devices (IMDs), and emergency shut-off mechanisms. These systems work together to detect and mitigate potential electrical hazards, providing multiple layers of protection for personnel and ensuring compliance with safety regulations and standards.

REFERENCES

[1]. Society of Automotive Engineers and USCAR, "SAE/USCAR-23 REVISION 1 – Road Vehicles – 60 V and 600 V Single Core (ISO/METRIC) Cables – Dimensions, Test Methods and Requirements," SAE Publishing, Warrendale, 2017.

[2]. Society of Automotive Engineers, "J3068 Identification of Vehicles and Supply Equipment for Conductive AC Charging," SAE Publishing, Warrendale, 2023.

[3]. Society of Automotive Engineers, "J3073 Battery Thermal Management," SAE Publishing, Warrendale, 2016.

[4]. L. Brabetz, M. Ayeb and L. Gysen, "Model-Based Circuit Protection Using Solid State Switches," SAE Int. J. Engines, vol. 10, no. 3, pp. 1375-1382, 2017, https://doi.org/10.4271/2017-01-1641.

[5]. Society of Automotive Engineers, "J2840 High Voltage Shielded and Jacketed Cable," SAE Publishing, Warrendale, 2021.

[6]. Society of Automotive Engineers, "J2344 Guidelines for Electric Vehicle Safety," SAE Publishing, Warrendale, 2020.

[7]. Society of Automotive Engineers, "J1673 High Voltage Automotive Wiring Assembly Design," SAE Publishing, Warrendale, 2012.

[8]. Society of Automotive Engineers, "J1715 Hybrid Electric Vehicle (HEV) and Electric Vehicle (EV) Terminology," SAE Publishing, Warrendale, 2022.

[9]. Society of Automotive Engineers, "J553 Circuit Breakers," SAE Publishing, Warrendale, 2022.

[10]. Z. Wu, X. Su, Y. Zhu and M. Xiao, "DC Link Capacitor Active Discharge by IGBT Weak Short Circuit," SAE Int. J. Adv. & Curr. Prac. in Mobility, vol. 1, no. 3, pp. 1177-1187, 2019, https://doi.org/10.4271/2019-01-0606.

[11]. Society of Automotive Engineers, "J2183 60 V and 600 V Single-Core Cables," SAE Publishing, Warrendale, 2012.

I

IEC 62196

IEC 62196, commonly called the "Type 2" standard, is a set of international standards developed by the IEC for EV conductive charging systems. It specifies the configurations and general requirements for the vehicle plugs, socket outlets, vehicle connectors, and vehicle inlets used in EVs and HVs. The standard ensures interoperability and safety across different EV charging systems globally.

Key Features:

- **Types of Connectors:** IEC 62196 includes several connector types, such as Type 1 (mostly used in North America and Japan), Type 2 (common in Europe), and others tailored for specific regional standards.
- **Charging Modes:** It outlines various charging modes, accommodating both AC and DC charging, to support a wide range of EVs and charging infrastructure.
- **Safety Provisions:** The standard includes critical safety features such as mechanisms for confirming the connection status, ensuring proper grounding, and protecting against electric shocks.
- **Communication Protocols:** It specifies the communication interfaces between the vehicle and the charging station, facilitating smart charging features, including charge management and billing.

IEC 62196 is pivotal for manufacturers of EVs, charging stations, and electrical components who aim to meet international safety and compatibility standards. It is also essential for governments and regulatory bodies to design and implement EV infrastructure to support environmental policies and reduce carbon emissions.

Inductive Charging System

The inductive charging system consists of two main components: a charging pad or base installed on the ground (charging station) and a receiving pad or coil mounted on the vehicle's underside (**Figure I.1**). When the EV or PHEV is positioned over the charging pad, electromagnetic induction occurs between the two coils, creating a magnetic field that transfers electrical energy wirelessly. The charging pad transfers AC electricity, which is then transmitted to the vehicle's coil and converted back DC to charge the battery.

FIGURE I.1 Wireless power transfer flow block diagram [1].

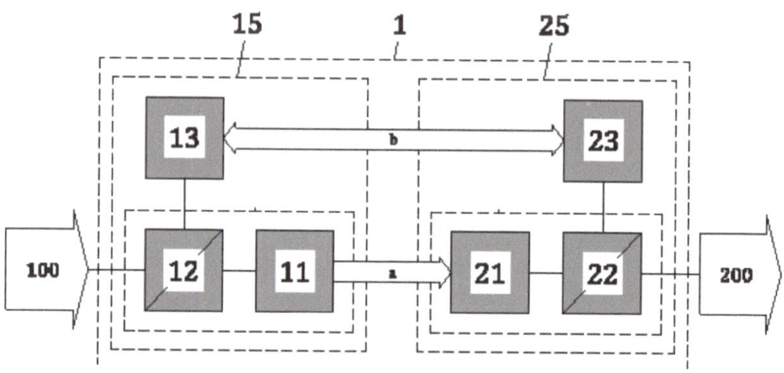

Key			
1	WPT system	21	VA coil
11	GA coil	22	VA electronics
12	GA electronics	23	Vehicle communication controller
13	Base communication controller	25	VA - vehicle subsystem
15	GA - ground subsystem	200	Power to battery
100	Power from grid	b	Communication
a	Wireless power flow		

Input Speed

In EVs, input speed is a critical parameter for the motor controller, which adjusts the electrical input to the motor to control the speed and torque output according to the vehicle's demand. The input speed in EVs is often governed by the power electronic systems designed to optimize energy usage and extend the battery life.

In HVs, which combine an ICE with an electric motor, input speed becomes a significant factor in managing the integration between the two power sources. The control system in hybrids uses input speed data to decide the optimal points for switching between electric power and combustion power, enhancing fuel efficiency and reducing emissions.

Input Split

In EVs and HVs, the input split refers to the system design where the power flow from the engine and electric motor is combined before being transmitted to the drivetrain. This setup is part of what is known as a "series–parallel hybrid" configuration, allowing the vehicle to operate on electric power, ICE power, or a combination of both, depending on driving conditions and battery charge level.

The input split architecture enables optimal power management and efficiency by allowing the vehicle's control system to choose the most efficient mode of operation. It combines the strengths of both power sources, utilizing the electric motor for low-speed operations and assistance during peak power demands. At the same time, the engine can take over at higher speeds or when higher power output is needed.

Insertion Force (N)

The force required to fully insert a set of mating components.

For more information, see **J1715** [1].

Insulator

An insulator in EVs and HVs is a material used to prevent the unwanted flow of electricity. Its primary function is to isolate electrical components, conductors, and other parts from each other and the vehicle's chassis, thus ensuring safety, functionality, and efficiency.

In EVs and HVs, insulators are crucial for the following:

- **Safety:** They prevent electrical shocks and protect passengers by insulating high-voltage components.
- **Performance:** By preventing electrical leaks, insulators help maintain the efficiency and performance of the vehicle's electrical system.
- **Durability:** Insulators protect sensitive components from environmental factors, such as moisture and heat, thereby extending the life of these components.

Common materials for insulators in these vehicles include the following:

- **Rubber:** Rubber is flexible and durable and is often used for wiring insulation and protective coverings.
- **Plastics:** Various plastics, such as polyethylene and polyvinyl chloride (PVC), are used due to their excellent insulative properties and resistance to degradation.
- **Ceramics:** Ceramics are used in high-temperature environments or where greater physical strength is needed.
- **Composite Materials:** These are increasingly used for their superior strength-to-weight ratios and excellent insulation characteristics.

Integrated Motor Assist

See **Motor**.

Interlock

An interlock in EVs and HVs refers to a safety mechanism designed to prevent the operation of certain vehicle functions unless specific conditions are met. This feature is crucial in managing the high-voltage electrical systems that power these vehicles, ensuring the safety of occupants and service personnel.

The primary function of an interlock is to avoid electrical hazards by ensuring that high-voltage components are not live when they are being serviced or when safety-critical conditions are not met. For instance, an interlock might prevent the vehicle from starting if the charging cable is connected or if a maintenance cover is removed. It also ensures that the electric drive system is deactivated during a crash.

Interlocks are integrated into various vehicle parts, including the battery pack, charging system, and high-voltage connectors. They typically involve a combination of physical switches and software controls that communicate with the vehicle's central control unit and ensure an unsafe combination of stimuli or events.

Vehicles must comply with specific international standards concerning electrical safety, such as the **ISO 6469** series or **SAE J2578** [2].

Interlock Systems

The HVIL is a system intended to protect people from exposure to high voltage or other hazardous conditions. It typically detects unwanted access or faults by passing a small (non-hazardous) signal through a loop connecting a set of normally closed conductors, connectors, sensors, and switches to check for electrical continuity [2].

Interoperability Class 1 GA (WPT)

Interoperability Class I GA (WPT) is a classification within EV and HV technology that specifies the standards and protocols for wireless charging systems. This classification ensures compatibility between wireless charging stations and a diverse range of EVs and HVs.

Interoperability Class I GA focuses on facilitating the seamless transfer of electrical power from a charging station to a vehicle without needing physical connectors or cables. It is part of a broader effort to standardize WPT technologies to enhance user convenience and safety and to promote the adoption of EVs by eliminating some of the hurdles associated with wired charging methods.

Typically designed for lower power applications, Class I WPT systems are generally used for light EVs such as scooters, motorcycles, and small personal mobility devices. These systems usually handle power levels up to a few kilowatts.

See also **WPT (Wireless Power Transfer)**.

Interoperability Class II GA (WPT)

Interoperability Class II GA (WPT) is a standardized category within the wireless charging technology used for EVs and HVs. This class designates explicitly the systems and protocols that enable wireless electrical power transmission from a charging station to the vehicle without physical connectors or cables.

These systems are intended for higher power demands and are suitable for passenger vehicles such as cars and light commercial vehicles. Class II can handle significantly higher power levels, ranging from several kilowatts up to tens of kilowatts.

See also **WPT (Wireless Power Transfer)**.

Inverter

An inverter is an essential electronic device that converts DC from the vehicle's battery pack into AC. This AC powers the electric traction motor, which drives the vehicle's wheels.

The inverter plays a pivotal role in the vehicle's powertrain by managing the electricity flow from the battery to the motor. It adjusts the AC output's frequency and amplitude to control the electric motor's speed and torque. In HVs, the inverter manages the power interchange between the electric motor and the ICE.

Bi-directional inverter is a term used for a device that can convert from AC to DC in one direction to serve as a battery charger and then be capable of being reversed and converted from DC to AC in the other direction to serve as an inverter [3].

An inverter consists of several key components, including power electronic switches (such as IGBTs or MOSFETs), capacitors, a cooling system, and control electronics that manage the conversion process through sophisticated software algorithms.

An inverter's efficiency, durability, and performance directly influence the vehicle's overall efficiency, driving range, and reliability (Figure I.2). Advanced inverters aim to minimize energy losses during the conversion process, enhance the motor's responsiveness, and operate reliably under varying environmental conditions.

FIGURE I.2 Example of inverter system models [3].

Inverter System Model Number		
A141J4aaa	A141J4bbb	A141J4ccc

Top Assembly	Inverter System Models						
	xxxG01			xxxG02		xxxG03	
Revision	-	A	B	-	A	-	A

	Description	Core Inverter System Configurations							
Core Inverter System	Part A	A1	A1	A1	A1	A1	A2	A2	Inverter System
	Part B	B1	B1	B1	B1	B1	B1	B1	
	Part C	C1	C2	C2	C2	C3	C3	C4	
	Part D	D1	D1	D2	D2	D2	D2	D2	
	Part E	E1	E1	E1	E2	E2	E2	E2	
	Part F Part G ... Part ZZ	COMPONENTS NOT SELECTED BY VM TO BE PART OF INVERTER SYSTEM MODEL							

NOTE: All Changes are Class I at Part Level
C2 are Class II at Top Assembly Level (Rev Letter)
E2 are Class I at Top Assembly Level (Group Change)

IP64 (Ingress Protection-64)

IP64 is a rating that denotes the level of protection an EV's or HV's electrical enclosures offer against the intrusion of solid objects and water. This rating is crucial for assessing the durability and safety of components that are exposed to environmental elements.

The IP rating system is composed of two digits. The first digit, "6," indicates complete protection against dust and other particulates, meaning no dust ingress occurs, which is essential for maintaining the integrity and functionality of electrical systems. The second digit, "4," refers to the enclosure's ability to protect against water splashes from all directions. Although it offers less protection against water compared to higher ratings, it is generally sufficient for protecting components in most driving conditions encountered by EVs and HVs.

In EVs and HVs, the IP64 rating is typically applied to components such as battery packs, connectors, and OBCs. These parts require robust protection to ensure long-term reliability and performance, especially considering the sensitivity of electrical systems to contaminants like dust and moisture.

Iron Losses

Iron losses, also known as core losses or magnetic losses, refer to the energy dissipated as heat in the magnetic cores of the electric motor used in EVs and HVs. These losses occur primarily due to hysteresis and eddy currents, which arise when the magnetic material within the motor's core is subjected to alternating magnetic fields during operation.

- **Hysteresis Losses:** These losses result from the magnetic induction lagging behind the magnetizing force within the core material. Each time the magnetic field is reversed, a certain amount of energy, proportional to the area of the magnetic material's hysteresis loop, is dissipated as heat. The frequency of the magnetic reversals and the properties of the core material (such as coercivity and remanence) significantly influence hysteresis losses.
- **Eddy Current Losses:** These are caused by circulating currents induced in the conductive material of the motor core by the changing magnetic field. The magnitude of eddy currents, and thus the losses, depends on the material's conductivity, the frequency of the magnetic field changes, and the thickness of the core plates. Reducing the thickness of the core plates and insulating them from each other can effectively minimize these losses.

Iron losses contribute to overall energy inefficiency in EVs and HVs, affecting the performance and efficiency of electrical components such as motors and transformers. The heat generated due to iron losses necessitates cooling mechanisms to prevent overheating and ensure the longevity of components. Additionally, minimizing iron losses through high-quality magnetic materials, efficient core designs, and advanced insulation techniques is crucial for optimizing EVs' and HVs' energy efficiency and range.

ISC (Infrastructure-Side Connection)

The connections made for the ACD system on the infrastructure side of the connection. The connection is made when the infrastructure-side connection mates with the vehicle-side connection. The ISC power transfer is analogous to the term "EVSE" which is applicable to conductive power transfer systems.

For more information, see **J1715** [1].

REFERENCES

[1]. Society of Automotive Engineers, "J1715 Hybrid Electric Vehicle (HEV) and Electric Vehicle (EV) Terminology," SAE Publishing, Warrendale, 2022.

[2]. Society of Automotive Engineers, "J2578 Recommended Practice for General Fuel Cell Vehicle Safety," SAE Publishing, Warrendale, 2023.

[3]. Society of Automotive Engineers, "J3072 Interconnection Requirements for Onboard, Grid Support Inverter Systems," SAE Publishing, Warrendale, 2021.

J

JEVS (Japan Electric Vehicle Association Standards) G105-1993

JEVS G105-1993 is a technical standard for EVs and HEVs that specifically focuses on measuring power consumption and energy efficiency. JEVS G105-1993, also known as "Energy Consumption Measurement Method for Electric Vehicles," outlines the procedures and criteria for evaluating the energy consumption of electric propulsion systems in vehicles.

The standard defines the methodology for measuring the energy consumption of electric propulsion systems during various driving cycles and operating conditions [1, 2]. This includes procedures for measuring energy consumption at different speeds, accelerations, and loads to simulate real-world driving scenarios.

JEVS G105-1993 specifies standardized driving cycles or test procedures that EVs must undergo to assess their energy efficiency. These driving cycles may include urban, highway, or a combination, allowing for comprehensive energy consumption measurements.

The standard establishes metrics and parameters for quantifying energy consumption and efficiency, such as energy consumption per unit distance traveled (e.g., watt-hours, Wh/km or kWh/100 km). These metrics provide valuable insights into the vehicle's overall energy performance and efficiency.

JEVS G105-1993 defines the test conditions, environmental factors, and equipment requirements necessary for conducting energy consumption measurements accurately and consistently. This ensures that test results are reliable, reproducible, and comparable across different EV models.

The standard outlines procedures for collecting, recording, and analyzing energy consumption data obtained during testing. This includes calculating average energy consumption, energy recovery during regenerative braking, and other relevant performance indicators.

REFERENCES

[1]. US Federal Government, "Emission Standards Reference Guide - EPA Urban Dynamometer Driving Schedule (UDDS)," April 18, 2024, accessed June 7, 2024, https://www.epa.gov/emission-standards-reference-guide/epa-urban-dynamometer-driving-schedule-udds.

[2]. Bielaczyc, P., Szczotka, A., and Woodburn, J., "Carbon Dioxide Emissions and Fuel Consumption from Passenger Cars Tested Over the NEDC and WLTC – An Overview and Experimental Results from Market Representative Vehicles," *2nd International Conference on the Sustainable Energy and Environmental Development*, Bielsko-Biala, Poland, 2019.

K

kW (Kilowatt)

A unit of power that is equivalent to 1000 W. It measures the rate of energy conversion.

$$\text{Power}(\text{kW}) = \left(\frac{\text{Energy}(\text{J})}{\text{Time}(\text{sec})}\right) \times 10^{-3}$$

where

Energy is measured in joules (J)
Time is measured in seconds (sec)
The term 10^{-3} converts the power from watts (W) to kilowatts (kW)

kW - Charging

kW charging refers to the rate at which electric energy is transferred from a charging station to an EV battery. It is measured in kW, indicating the charger's power level. The higher the kW, the faster the charging speed.

If an EV uses a charger rated at 10 kW for 2.5 h, the energy consumed can be calculated as follows:

$$\text{kWh} = 10\,\text{kW} \times 2.5\,\text{h} = 25\,\text{kWh}$$

kW - Vehicles

An EV's battery's total energy storage capacity is expressed in kWh. This figure is pivotal in determining the vehicle's range; higher kWh values typically indicate a longer driving range on a single charge.

EVs also measure energy consumption in kilowatt-hours per 100 kilometers (kWh/100 km) or per mile (kWh/mi). This measurement helps users understand how efficiently a vehicle uses electricity.

kWh (Kilowatt-Hour)

A kWh is a unit of energy measurement equivalent to 1 kW of power expended for 1 h. It is used to denote EVs' energy capacity and consumption.

L

LD-PEV (Light-Duty Plug-In Vehicle)

An LD-PEV is a subclass of EVs designed primarily for personal transportation and commercial uses that fall under light-duty classifications. These vehicles are equipped with rechargeable battery packs that can be plugged into an external electric power source. The category encompasses various types of vehicles including all-electric vehicles, BEVs, PHEVs, and EREVs.

Level 1 Charging (AC)

Level 1 charging refers to the primary method of charging EVs and PHEVs using a standard AC household outlet. This level of charging is typically the slowest among different charging levels but provides a convenient and accessible option for EV owners who have access to a standard electrical outlet.

Level 1 charging utilizes a standard 120-V AC outlet in residential buildings and workplaces. The charging power output for Level 1 charging is typically around 1.4 to 1.9 kW. However, the actual charging rate can vary depending on the vehicle's OBC and the quality of the electrical supply (Table L.1).

TABLE L.1 Charge level specification (North America) [1].

Charge method	Nominal supply voltage	Max current (Amps-continuous)	Branch circuit breaker rating (Amps)	Continuous input power (reference)
Level 1	120 VAC, 1-phase	12 A	15 A	1.44 kW
Level 2	208–240 VAC, 1-phase	32 A	40 A	6.66–7.68 kW
Level 3	208–600 VAC, 3-phase	400 A	As required	>7.68 kW

Level 2 Charging (AC)

Level 2 charging refers to a faster and more robust method of charging EVs and PHEVs using an AC charging station with higher voltage and amperage compared to Level 1 charging. It is commonly used in residential, commercial, and public charging infrastructures to provide quicker charging times for EV owners.

Level 2 charging utilizes a 240-V AC power supply in residential garages, workplaces, parking lots, and public charging stations. The charging power output can range from 3.3 to 19.2 kW or higher depending on the charging station's capabilities and the vehicle's OBC.

Level 2 EVSE

Level 2 EVSE refers to a charging infrastructure designed to supply electric power to EVs and PHEVs at a higher charging rate than Level 1 chargers. Level 2 EVSE operates on AC and is commonly installed in residential, commercial, and public locations to facilitate faster charging of EV batteries.

Level 2 EVSEs provide a faster charging rate than Level 1 chargers, typically delivering around 10 to 60 mi of range per hour of charging, depending on the vehicle and the EVSE's power output. Level 2 chargers operate at a higher voltage of 208 to 240 V AC and can deliver higher amperage, often ranging from 16 to 80 amps, allowing for faster charging times.

Level 2 EVSE uses connectors such as the SAE J1772 plug for North America or the Type 2 plug for Europe and other regions, providing a standardized interface for EV charging. It requires a dedicated electrical circuit and may involve professional installation, including the mounting of charging stations, electrical wiring, and appropriate safety measures. Most EVs and PHEVs can use Level 2 EVSE, making it a versatile charging solution for various EV models and brands.

Level 2 EVSE significantly reduces charging time compared to Level 1 chargers, making it suitable for EV owners who require quicker turnaround times or need to charge their vehicles more frequently. With Level 2 EVSE installed at home, EV owners can charge their vehicles overnight or during off-peak hours, maximizing convenience and minimizing downtime. Level 2 chargers are commonly installed in public locations such as shopping centers, workplaces, and parking facilities, providing EV owners access to faster charging options while on the go.

Level 3 Charging (DC)

Level 3 charging, also known as DC fast charging or rapid charging, is a high-power charging method for EVs and PHEVs that utilizes DC to recharge the vehicle's battery rapidly. Level 3 charging stations can deliver significantly higher power output than Level 1 and Level 2 chargers, resulting in much faster charging times.

Level 3 charging stations operate at high voltages and currents, typically 200 to 800 V or more, with power outputs exceeding 50 kW and often reaching 350 kW or higher. These high-power charging stations are designed to quickly replenish an EV's battery, making them suitable for long-distance travel and reducing charging downtime.

$LiCoO_2$ (Lithium Cobalt Oxide)

$LiCoO_2$, also known as LCO, is a Li-ion battery cathode material. It is widely recognized for its high energy density and is primarily utilized in consumer electronics such as smartphones, laptops, and digital cameras. Though less predominant in EV battery applications due to its relatively short life span and issues with thermal stability, it remains a foundational material in the study and development of advanced lithium-based batteries. Research continues to enhance its safety and longevity to broaden its usability in the EV industry.

Life Cycle Testing of EV Battery Modules

Life cycle testing of EV battery modules refers to the comprehensive examination and evaluation of battery modules throughout their operational lifespan. This process involves a series of tests designed to simulate real-world conditions and stresses the battery will encounter over its expected life. The objective is to assess the battery modules' performance, durability, and reliability, ensuring they meet or exceed specified standards and requirements for EVs. Testing consists of environmental, mechanical, electrical, and safety elements.

For additional information, see **J2288**, **J1715**, and **J1797** [2–4].

$LiFePO_4$ (Lithium Iron Phosphate Batteries)

$LiFePO_4$ batteries are Li-ion batteries that use lithium iron phosphate as the cathode material (Figure L.1). Characterized by their long cycle life and stability, these batteries offer significant advantages in safety due to their high thermal threshold and resistance to thermal runaway.

The battery comprises a lithium iron phosphate cathode, a graphite anode, and typically a lithium salt in an organic solvent as the electrolyte. The iron phosphate structure contributes to the battery's robustness and safety.

FIGURE L.1 Model of the molecular layout of lithium iron phosphate.

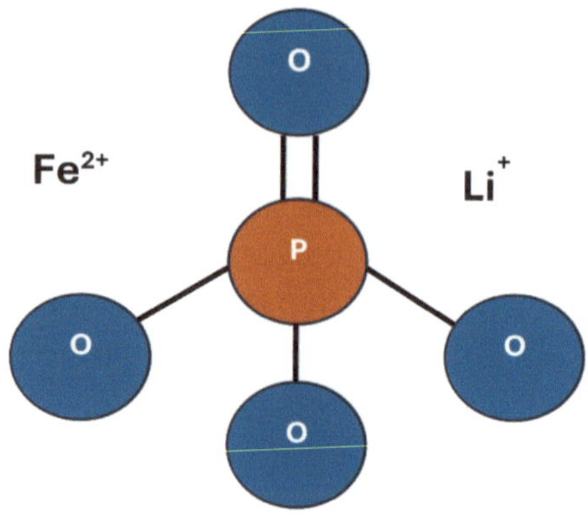

Li-ion (Lithium-Ion) Battery

A Li-ion battery is a rechargeable battery widely used in EVs and HEVs due to its high energy density, lightweight construction, and long cycle life (Figure L.2). Li-ion batteries are composed of lithium compounds as electrolytes, facilitating the movement of lithium ions between the battery's positive (cathode) and negative (anode) electrodes during charging and discharging cycles.

FIGURE L.2 Comparison of attributes of various ESSs [5].

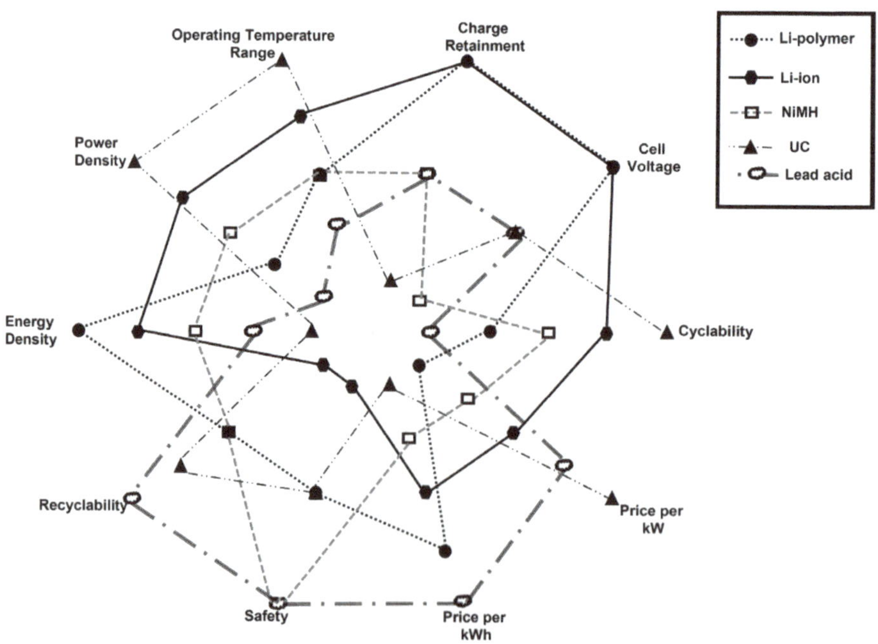

The cathode material, typically lithium cobalt oxide ($LiCoO_2$), lithium manganese oxide ($LiMn_2O_4$), or lithium iron phosphate ($LiFePO_4$), determines the battery's energy density and performance characteristics. The anode commonly comprises graphite, efficiently absorbing and releasing lithium ions during charging and discharging.

Li-ion batteries use a nonaqueous electrolyte solution containing lithium salts, such as lithium hexafluorophosphate ($LiPF_6$), to facilitate ion movement between electrodes. A thin, porous membrane separates the cathode and anode, preventing short circuits while allowing ion transport.

Low-Voltage DC–DC Converter

A low-voltage DC–DC converter is an essential component in EVs and HVs that facilitates the conversion of electrical power from a high-voltage level to a low-voltage level (Figure L.3). This converter plays a crucial role in managing the power distribution within the vehicle's electrical system (sensors and ECUs), ensuring the efficient operation of various subsystems and components.

FIGURE L.3 An example of the vehicle's high- and low-voltage systems.

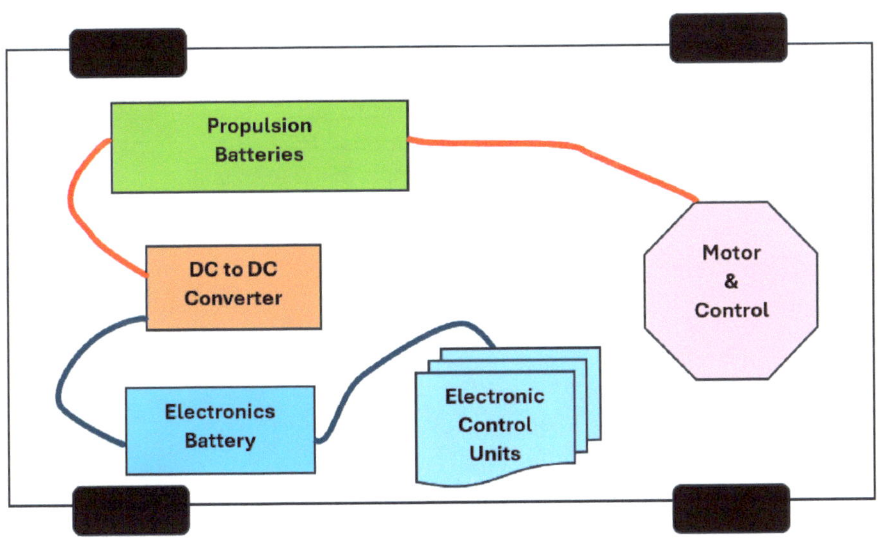

Low-Voltage System

A low-voltage system in EVs refers to the electrical circuits and components that operate at a voltage level typically below 60 V DC or 30 V AC. This system is distinct from the high-voltage components that drive the electric motor and handle propulsion.

The low-voltage system powers the vehicle's auxiliary functions, such as lighting, infotainment, climate control, and ECUs. These systems are often powered by a conventional 12-V battery, like those in ICE vehicles.

REFERENCES

[1]. Society of Automotive Engineers, "J1773 SAE Electric Vehicle Inductively Coupled Charging," SAE Publishing, Warrendale, 2014.

[2]. Society of Automotive Engineers, "J2288 Life Cycle Testing of Electric Vehicle Battery Modules," SAE Publishing, Warrendale, 2020.

[3]. Society of Automotive Engineers, "J1715 Hybrid Electric Vehicle (HEV) and Electric Vehicle (EV) Terminology," SAE Publishing, Warrendale, 2022.

[4]. Society of Automotive Engineers, "J1797 Recommended Practice for Packaging of Electric Vehicle Battery Modules," SAE Publishing, Warrendale, 2016.

[5]. S. M. Lukic, J. Cao, R. C. Bansal, F. Rodriquez and A. Emadi, "Energy Storage Systems for Automotive Applications," *IEEE Transactions on Automotive Applications*, vol. 55, no. 6, pp. 2258-2267, 2008.

M

Manual Disconnects

Manual disconnects, also known as manual service disconnects or emergency shut-off switches, are safety devices installed in EVs and HEVs to disconnect the high-voltage electrical system from the vehicle's components and circuits [1]. These devices are designed to quickly and effectively isolate the vehicle's electrical power during emergencies, maintenance tasks, or servicing procedures [2].

Manual disconnects serve as a safety mechanism to protect individuals working on or around EVs and HEVs from potential electrical hazards. When activated, the manual disconnect interrupts the flow of electricity from the high-voltage battery to the vehicle's propulsion system, electrical components, and auxiliary systems, ensuring that no electrical power is present during maintenance or emergencies.

Mechanical Point or Mechanical Ratio

In the context of EVs and HVs, a mechanical point or mechanical ratio refers to the relationship between the rotational speed of the electric motor(s) and the rotational speed of the wheels. This concept is critical in the design and functionality of the drivetrain system, which includes components like the gearbox and differential.

The mechanical ratio is essential for optimizing the performance and efficiency of EVs and HVs. It determines how the torque produced by the motor is transmitted to the wheels, affecting the vehicle's acceleration, top speed, and energy consumption. Adjusting the mechanical ratio can balance between high-speed efficiency and low-speed torque.

Key Features:

1. **Transmission Efficiency:** A well-optimized mechanical ratio ensures that the electric motor operates within its most efficient range, improving overall vehicle efficiency.
2. **Torque Multiplication:** The mechanical ratio multiplies the torque from the motor before it reaches the wheels, crucial for providing sufficient acceleration from a standstill.

3. **Speed Adaptation:** It allows the vehicle to maintain an optimal speed by adjusting the output according to the motor's revolutions per minute (RPM) and the vehicle's current speed requirements.
4. **Simplicity and Reliability:** In many EVs, the mechanical ratio is fixed, simplifying the transmission system and reducing maintenance compared to traditional vehicles with complex multi-gear transmissions.

Applications:

- **Single-Speed Transmission:** Most EVs use a single-speed gearbox with a fixed mechanical ratio, designed to balance acceleration and top speed while keeping the system simple and lightweight.
- **Variable-Speed Transmission:** Some high-performance models and HVs might use more complex transmission systems with variable mechanical ratios to optimize performance across a broader range of conditions.

Microgrid

A microgrid is a localized group of electricity sources and loads that normally operates connected to and synchronously with the traditional centralized electrical grid (microgrid) but can disconnect and function autonomously as physical and economic conditions dictate. In the context of EVs and HVs, microgrids facilitate the integration of vehicle charging infrastructure with localized renewable energy sources, enhancing energy sustainability and resilience.

Microgrids for EVs and HVs incorporate various DERs like solar panels, wind turbines, and battery storage systems, alongside vehicle charging stations. This setup not only supports the charging of EVs and HEVs but also ensures that the charging process is managed in an efficient, eco-friendly manner. Microgrids can operate independently or in conjunction with the main grid, providing flexibility and security to the energy supply, especially in areas prone to outages or with weak grid infrastructure.

1. **Energy Resilience:** Microgrids can maintain operation independently of the main grid, providing reliable power during outages and reducing downtime for vehicle charging.
2. **Renewable Integration:** They facilitate the increased use of renewable energy sources for EV and HEV charging, reducing the carbon footprint associated with vehicle operations.
3. **Load Management:** Microgrids can dynamically manage the load from vehicle charging stations, optimizing energy use and preventing overloads on the local grid infrastructure.

4. **Cost Efficiency:** By using locally generated power, especially during peak tariff periods, microgrids can reduce energy costs associated with vehicle charging.
5. **Smart Charging:** Advanced microgrids utilize smart meters and Internet of Things (IoT) devices to optimize charging schedules based on energy availability and vehicle usage patterns.

Microgrids are particularly beneficial in remote or isolated communities, commercial campuses, and residential complexes with high concentrations of EVs and HEVs. They are also vital in disaster-prone areas, providing an uninterrupted power supply when the main grid fails.

Mi/kWh (Miles per Kilowatt-Hour)

Mi/kWh is a unit of measurement used to express the energy efficiency of EVs and HEVs. It indicates the miles a vehicle can travel on 1 kW-h of electricity. This metric is analogous to miles per gallon (mpg) in conventional ICE vehicles but specific to electrical energy consumption.

Mi/kWh is a crucial efficiency standard for assessing the performance of EVs and HVs. It provides consumers with a clear indicator of a vehicle's range and efficiency, which are key factors in the decision-making process for potential EV and HEV owners. Higher mi/kWh values signify greater efficiency, as the vehicle can travel further on less electrical energy.

$$\text{Miles per Kilowatt} - \text{Hour} = \frac{\text{Distance traveled in miles}}{\text{Energy used in kWh}}$$

Mild HV (Hybrid Vehicle)

A mild HV, also known as a power-assist hybrid, is an HEV that incorporates a small electric motor and battery system to assist the ICE in certain driving conditions. Unlike full-HVs or PHEVs, mild HVs cannot operate solely on electric power. Instead, they utilize the electric motor to provide additional power during acceleration, reduce engine load, and enable engine start–stop functionality to enhance fuel efficiency and reduce emissions.

Mild hybrid technology has become increasingly popular among automakers seeking to improve their vehicles' fuel efficiency and environmental performance without the cost and complexity of complete hybrid systems.

Molten Salt Battery

A molten salt battery is a type of battery that utilizes molten salts as an electrolyte. It operates at high temperatures where the salt is in a liquid state, facilitating the movement of ions between electrodes to produce electrical energy. This technology is particularly noted for its high energy density and long cycle life, making it suitable for EVs and HEVs.

Molten salt batteries, often comprising sodium–sulfur or lithium–sulfur combinations, function at temperatures ranging from 300 to 700°C (572 to 1292°F). These batteries are known for storing large amounts of energy, which is essential for the extended-range requirements of EVs and HEVs. They are also capable of rapid charging and discharging, a critical feature for automotive applications.

Motor

An EV motor is used to drive EVs and HEVs. Unlike ICEs in traditional vehicles, electric motors convert electrical energy into mechanical energy, providing propulsion.

Electric motors are essential components in the powertrain of EVs and HEVs, offering high efficiency, excellent torque characteristics, and smooth operation. These motors can be powered by various sources, including batteries, hybrid systems, or hydrogen fuel cells. The design and type of motor can vary widely depending on the vehicle's specific requirements, including performance, range, and cost considerations.

Key Types:

1. **DC Motors:** DC motors were among the first used in EVs due to their simplicity and low cost. However, they are less efficient and have largely been supplanted by more advanced motors in modern EVs.
2. **Induction Motors:** Popularized by some major EV manufacturers, induction motors are robust, reliable, and capable of operating without rare earth metals, making them cost-effective and durable.
3. **PMSM:** These motors are known for their high efficiency and power density, resulting from using permanent magnets in the rotor. They are widely used in both consumer and high-performance EVs.
4. **SRMs:** Known for their simplicity and durability, SRMs are gaining popularity due to not requiring permanent magnets and thus being less affected by fluctuating rare earth material costs.

Motor Drive

A motor drive in the context of EVs and HVs refers to the complex system that controls the electric motor's operation, including its speed, torque, and direction. This system is crucial for ensuring the efficient and effective propulsion of the vehicle.

The motor drive system consists primarily of a power inverter and a controller. The power inverter converts DC power from the vehicle's battery into AC power to drive the electric motor, while the controller regulates the inverter's output based on inputs from the vehicle's accelerator pedal, speed sensors, and other controls. This integration allows for precise control over the motor's performance, optimizing the vehicle's responsiveness and energy efficiency.

Motor Generators

A motor generator in EVs and HVs refers to a device that serves dual roles: as an electric motor that drives the vehicle and as a generator that converts mechanical energy into electrical energy. This integral component is key in managing power delivery and energy recuperation in these vehicles.

The motor generator is a critical drivetrain element in EVs and HEVs. It functions primarily as an electric motor to provide propulsion. In hybrids, it also acts in concert with an ICE to optimize efficiency and performance. As a generator, it harnesses kinetic energy during braking or coasting phases (regenerative braking) and converts it back into electrical energy, which is then stored in the vehicle's battery system.

1. **Dual Functionality:** Operates as both a propulsion motor and a generator for energy recuperation.
2. **Regenerative Braking:** Converts kinetic energy into electrical energy during braking, enhancing the vehicle's overall efficiency and extending the driving range.
3. **Energy Efficiency:** Improves fuel economy in hybrids by supporting the ICE and reducing reliance on fossil fuels.
4. **Seamless Integration:** Typically integrated with the vehicle's transmission system to ensure smooth power delivery.
5. **Variable Speed Control:** Equipped with advanced control systems that manage the speed and torque of the motor for optimal performance and energy usage.

Motor generators are used across all types of electrically powered or assisted vehicles, including:

- **Pure EVs:** Where they serve as the primary source of vehicle propulsion.
- **HEVs:** Where they assist an ICE and can drive the vehicle independently when needed.
- **PHEVs:** These vehicles can drive in full electric mode or as a hybrid, utilizing the motor generator in both capacities.

MPGe (Miles per Gallon Equivalent)

MPGe is a metric used to measure the energy efficiency of EVs and HVs in terms comparable to traditional ICE vehicles. It represents the distance an electric or hybrid car can travel on an amount of energy equivalent to 1 gal of gasoline, providing consumers with a standardized way to evaluate and compare fuel efficiency across different vehicle types.

The MPGe rating is calculated based on the energy content of a gallon of gasoline, which is approximately 33.7 kWh. For EVs and HVs, energy consumption is measured in kWh/100 mi (or another specified distance). By dividing the energy consumption by the energy content of a gallon of gasoline, the MPGe rating is derived, indicating how many miles an EV or HV can travel on the energy equivalent to 1 gal of gas.

MPGe provides consumers with a standardized unit of measurement that helps them compare the energy efficiency and operating costs of EVs and HVs against traditional gasoline-powered vehicles. It offers a clear and straightforward way to understand the fuel efficiency of electric and hybrid models, aiding consumers in making informed decisions based on their driving habits, energy costs, and environmental considerations. The energy content of 1 gal of gasoline is approximately 33.7 kWh. The MPGe is calculated using the following equation:

$$\text{MPGe} = \frac{\text{Distance in miles}}{\text{Energy used in kWh}} \times 33.7$$

REFERENCES

[1]. Society of Automotive Engineers, "J2344 Guidelines for Electric Vehicle Safety," SAE Publishing, Warrendale, 2020.

[2]. Society of Automotive Engineers, "J2990 Hybrid and Electric Vehicle Safety Systems Information Report," SAE Publishing, Warrendale, 2020.

NdFeB (Neodymium Iron Boron) Permanent Magnets

NdFeB permanent magnets are a type of rare earth magnet widely used in EVs and HVs. They are made from an alloy of neodymium, iron, and boron and are known for their exceptional magnetic strength and durability.

NdFeB magnets are crucial in the design of electric motors, particularly those used in EVs and HEVs. They are preferred for their high magnetic field strength and ability to maintain magnetism under high-temperature conditions, making them ideal for use in the compact, high-performance motors required for automotive applications.

- **High Magnetic Strength:** NdFeB magnets have a very high residual magnetic field compared to other types of magnets. This allows for the development of motors with higher efficiency and power density.
- **Thermal Stability:** These magnets can operate effectively at higher temperatures than many other types of permanent magnets, although they may require special coatings or stabilization techniques to maintain performance at peak temperatures.
- **Corrosion Resistance:** NdFeB magnets are prone to corrosion in harsh environments, which is why they are often coated with materials such as nickel, zinc, or epoxy to enhance durability.
- **Cost-Effectiveness:** While neodymium magnets are more expensive than ceramic or other types of ferrite magnets, their superior performance characteristics make them cost-effective for applications requiring strong, compact magnets.
- **Energy Efficiency:** The use of NdFeB magnets in vehicle motors helps to improve overall energy efficiency, reducing battery load and increasing vehicle range.

NEDC (New European Driving Cycle)

The NEDC is a standardized test procedure used to assess vehicles' fuel efficiency and emissions, including EVs and HEVs. Developed by the EU, the NEDC test cycle simulates typical urban and extra-urban driving conditions to evaluate a vehicle's performance in terms of energy consumption, range, and environmental impact.

The NEDC test cycle consists of four phases: urban driving, extra-urban driving, a combination of urban and extra-urban driving, and additional phases such as cold starts and engine idling. Each phase represents typical driving scenarios and traffic conditions encountered by vehicles on European roads.

The NEDC measures the fuel or energy consumed by a vehicle during the test cycle, expressed in liters per 100 kilometers (L/100 km) or kWh/100 km for EVs. The test cycle assesses the carbon dioxide (CO_2) emissions produced during the simulated driving scenarios, providing insights into the vehicle's environmental impact. By analyzing fuel consumption and energy usage, the NEDC evaluates vehicle energy efficiency, helping consumers and regulators understand the vehicle's performance in terms of energy conservation.

Net Ampere-Hours

For a discharge test including both negative (discharge) and positive (regen) current or power steps, the difference between the A·h removed from the battery during discharge steps and the A·h returned to the battery during regen steps, regardless of battery charge acceptance. Typical units are ampere-hour (A·h).

For more information, see **J1715** [1].

NEV (Neighborhood Electric Vehicle)

An NEV is a category of EV specifically designed for short-range, low-speed travel within residential areas, gated communities, and small towns. These vehicles are generally limited to speeds of about 25 to 35 mph and are legally restricted to roads with posted speed limits typically not exceeding 35 mph.

NEVs are compact, typically accommodating two to four passengers, and are powered solely by electric batteries. They are engineered to offer a convenient and eco-friendly transportation alternative for local commutes, errands, and community mobility. Due to their limited speed and range, NEVs are unsuitable for highways or high-speed travel.

NFC (Near-Field Communication) Tag

An NFC tag in the context of EVs and HVs is a small, wireless device that uses the NFC technology to facilitate data exchange between the vehicle and a charging station or other enabled devices without the need for direct contact or manual data entry [2].

NFC tags are embedded in EVs and HVs to streamline the process of authentication and payment at charging stations. By simply bringing the vehicle within a close range of an NFC-enabled charging station, the NFC Tag allows for automatic identification and initiation of the charging process, enhancing user convenience and efficiency.

NHEV-TSA (National Highway Electric Vehicle Traffic Safety Administration)

The NHEV-TSA is a regulatory agency tasked with overseeing and regulating safety standards and practices related to EVs operating on national highways. Unlike its counterpart, the National Highway Traffic Safety Administration (NHTSA), which focuses on conventional vehicles, the NHEV-TSA explicitly addresses the unique safety considerations associated with EVs.

NiCd (Nickel-Cadmium)

NiCd batteries are a type of rechargeable battery using nickel oxide hydroxide and metallic cadmium as electrodes. They have been used in EVs and HVs due to their robustness and good performance across a wide range of temperatures.

NiCd batteries were among the earliest battery types used in EVs and HEVs. Known for their durability and ability to deliver high power output, these batteries can endure extreme conditions. However, due to environmental concerns associated with cadmium, a toxic heavy metal, their use has diminished in favor of more environmentally friendly alternatives like Nickel Metal Hydride (NiMH) and Li-ion batteries.

- **Durability:** NiCd batteries are known for their long life and ability to withstand a high number of charge–discharge cycles.
- **Performance in Extreme Conditions:** They perform reliably under a wide range of temperature conditions, making them suitable for use in environments that experience extreme weather.
- **High Discharge Rate:** NiCd batteries can support high discharge rates without significant loss of capacity, which is beneficial for applications requiring bursts of power.
- **Environmental Impact:** The use of cadmium, a toxic heavy metal, poses environmental risks, particularly during disposal and recycling processes.

NiMH (Nickel-Metal Hydride)

NiMH refers to a type of rechargeable battery technology used in EVs and HEVs. NiMH batteries balance cost, performance, and environmental impact, making them popular for hybrid cars and other applications where battery weight and energy density are important considerations.

NiMH batteries were widely adopted in the early development of HVs due to their robustness, longer cycle life compared to NiCd batteries, and less environmental impact since they do not contain toxic heavy metals like cadmium. They consist of a positive electrode made from nickel oxyhydroxide (NiOOH) and a negative electrode made of a hydrogen-absorbing alloy.

- **Energy Density:** NiMH batteries typically have a higher energy density than lead-acid batteries but lower than Li-ion batteries. This makes them suitable for use where moderate energy density and weight considerations are important.
- **Cycle Life:** These batteries exhibit a good cycle life, which refers to the number of complete charge and discharge cycles a battery can support before its capacity falls under 80% of its original capacity.
- **Temperature Performance:** NiMH batteries perform well under a wide range of temperatures, making them suitable for use in diverse environmental conditions.
- **Cost-Effectiveness:** While not as cheap as lead-acid batteries, NiMH batteries are generally less expensive than Li-ion batteries, offering a good cost-to-performance ratio.
- **Environmental Impact:** NiMH batteries are more environmentally friendly compared to NiCd batteries, as they do not contain toxic metals and are easier to recycle.

NiZn (Nickel-Zinc)

An NiZn battery is a type of rechargeable battery that combines nickel oxide hydroxide and metallic zinc as electrodes. It offers a viable alternative to other battery chemistries for EVs and HEVs, providing a balance between cost, energy density, and environmental friendliness.

The NiZn battery chemistry is notable for its high energy density relative to its cost and environmental impact. It operates at a nominal voltage of 1.65 V, which is higher than lead-acid batteries and some other nickel-based batteries. This higher voltage can lead to fewer cells in series and potentially lower costs in battery pack construction. NiZn batteries are particularly praised for their ability to deliver high discharge rates and for their excellent performance in terms of cycle life when properly managed.

- **High Energy Density:** NiZn batteries offer higher energy density than traditional lead-acid batteries, making them suitable for applications where weight and space are critical constraints.
- **Cycle Life:** With proper management, NiZn batteries can achieve a competitive cycle life, though historically they have been prone to shorter lifespans due to dendrite formation during charging.
- **High Discharge Rate:** These batteries can handle high discharge rates, which is beneficial for acceleration in EVs.
- **Cost-Effectiveness:** They are generally less expensive than Li-ion batteries, making them a cost-effective solution for budget-conscious EV and HEV manufacturers.
- **Environmental Impact:** NiZn batteries use zinc, an abundant and less environmentally hazardous material compared to other metals used in battery technologies.

NiZn batteries are used in a variety of applications beyond EVs and HEVs, including emergency lighting, portable tools, and power storage. In the automotive sector, they are considered for both full-EVs and HVs where high power output and cost efficiency are required.

REFERENCES

[1]. Society of Automotive Engineers, "J1715 Hybrid Electric Vehicle (HEV) and Electric Vehicle (EV) Terminology," SAE Publishing, Warrendale, 2022.

[2]. Society of Automotive Engineers, "J2836 Use Cases for Diagnostic Communication for Plug-In Electric Vehicles," SAE Publishing, Warrendale, 2021.

O

OBC (Onboard Charger)

An OBC is an essential component in EVs and HEVs that converts AC from a power source, such as a domestic wall socket or public charging station, into DC to charge the vehicle's battery. This device is built into the vehicle and is critical for facilitating the battery charging process.

The OBC primarily manages the charging process, ensuring the battery is charged efficiently and safely. It regulates the voltage and current from the external power supply to match the battery's specifications. This optimizes the charging speed and helps prolong the battery's life by preventing overcharging and managing the battery's temperature during charging.

OBRESS (Onboard Rechargeable Energy Storage System)

An OBRESS refers to the system used in EVs and HVs to store the electrical energy necessary for the vehicle's operation. This system typically comprises a battery pack with one or more rechargeable battery modules. OBRESS is a critical component in the functionality of EVs and HEVs, providing the necessary power for propulsion and other electrical systems within the vehicle.

See also **Onboard EES (Electrical Energy Storage)**.

OCPP (Open Charge Point Protocol)

The OCPP is an application protocol for communication between EV charging stations and a central management system, commonly known as a charging station network. It is utilized globally to facilitate the networked operation of charging points, ensuring they can be monitored, maintained, and managed remotely and efficiently.

OCPP was developed to provide a standardized approach to the way charging equipment communicates with central systems, promoting interoperability and flexibility across different manufacturers and network providers. The protocol allows the implementation of various functionalities, such as remote start and stop of charging sessions, real-time monitoring, transaction handling, and firmware updates.

The protocol has several versions, with OCPP 1.6 and OCPP 2.0 being the most adopted. Each version enhances capabilities and security features, with OCPP 2.0 introducing improved security measures, more efficient communication, and additional features to support smarter energy management and integration with renewable energy sources.

For EV and HV owners, OCPP ensures a more reliable and accessible charging infrastructure. It allows EV charging service providers to offer better services, such as reservation systems for charging points, variable pricing models, and more user-friendly interfaces. For operators, it simplifies the management of large networks of chargers, optimizes the use of electrical resources, and enhances customer satisfaction through reliable and efficient services.

Onboard EES (Electrical Energy Storage)

Onboard electrical ESSs are integral components of EVs and HVs, designed to store electrical energy for propulsion. These systems primarily consist of batteries or other energy storage devices that accumulate and provide power to the vehicle's electric motor.

Types:

1. **Batteries:** The most common form of EES in vehicles, typically using Li-ion cells for their high energy density and efficiency. Other types include NiMH and lead-acid batteries.
2. **Supercapacitors (SCs):** Also known as ultracapacitors, store energy through a static charge as opposed to chemical reactions and are used alongside batteries to provide quick bursts of energy.
3. **Flywheel Energy Storage:** Mechanical devices that store energy in a rotating mass. Although less common, they are used for their ability to absorb and release energy quickly.

EES systems in EVs and HVs provide the necessary electrical power to the vehicle's motor, replacing or supplementing the ICE. In full-EVs, the EES is the sole source of propulsion power, while in hybrid configurations, it works in conjunction with an ICE, enhancing fuel efficiency and reducing emissions.

One-Pedal Driving

One-pedal driving is a driving mode available in many EVs and HVs that allows the driver to accelerate and decelerate using only the accelerator pedal. When the driver lifts their foot off the accelerator, the vehicle automatically initiates regenerative braking, which decelerates the car while simultaneously recharging the battery.

In traditional vehicles, the driver must use the brake pedal to decelerate. However, in vehicles equipped with one-pedal driving, the intensity of deceleration can be controlled by how much the driver eases off the accelerator pedal. Completely lifting the foot off the accelerator can bring the vehicle to a full stop in certain models.

One-pedal driving is especially advantageous in urban driving conditions where frequent stopping is required. The feature can also be turned on or off according to the driver's preference or driving conditions.

OpenADR (Open Automated Demand Response)

OpenADR is not an EV or HV but a communications standard that is crucial in integrating EVs within smart grid environments. Developed to facilitate DR and energy management solutions, OpenADR provides a framework for sending, receiving, and responding to signals designed to adjust power usage during peak periods or when the grid is under stress.

In the context of EVs and HVs, OpenADR can be instrumental in managing their charging behaviors to optimize energy use and costs. By integrating OpenADR protocols, EVs can communicate with utility providers to access real-time information on electricity rates, grid demands, and peak load times. This enables EVs to participate in DR events by adjusting charging times to off-peak periods, which can reduce charging costs for consumers and help balance the electrical load on the grid, enhancing overall grid reliability.

As the adoption of EVs continues to rise and the deployment of renewable energy sources grows, the role of OpenADR in managing these resources becomes increasingly important. Future advancements may lead to more sophisticated systems where EVs respond to grid demands and contribute back to the grid, further supporting the transition to a more sustainable and resilient energy ecosystem.

Operating Temperature

Operating temperature is the range of ambient temperatures within which EVs and HVs are designed to operate effectively and efficiently. This encompasses the temperature conditions under which the vehicle's battery, motor, and other critical systems function optimally.

Operating temperature is a critical parameter for EVs and HVs, primarily due to its impact on battery performance, efficiency, and lifespan. Batteries, typically Li-ion batteries, are sensitive to extreme temperatures. High temperatures can accelerate battery degradation, reduce capacity, and increase the risk of thermal events, while low temperatures can diminish battery efficiency and performance, increasing energy consumption and reducing vehicle range.

Optimized Charging

Optimized charging refers to using advanced algorithms and technologies to manage the charging process of EVs and HEVs to maximize efficiency, extend battery life, and minimize energy costs. This approach considers various factors, including battery health, charging rates, electricity grid demands, and user preferences, to determine the optimal times and rates for charging.

Output Split

Output split is a specific configuration used in EVs and HEVs to manage the distribution of power between the ICE and the electric motor. This system is part of a broader category of powertrain architectures known as power-split or series–parallel hybrids.

Key Characteristics:

- **Dual Power Sources:** Utilizes both an ICE and one or more electric motors. The output split configuration allows these sources to operate independently or in conjunction to provide propulsion.
- **Transmission Mechanics:** Typically involves a complex transmission system that combines the mechanical output from both power sources through a planetary gear set or similar mechanism.
- **Efficiency and Control:** Provides excellent control over the power delivery and the ratio of electric to gasoline power, enhancing fuel efficiency and reducing emissions.

In an output split system, the combustion engine can directly drive the vehicle, charge the batteries, or do both simultaneously. The electric motor can also add power to the driveline for acceleration or take over entirely for low-speed or low-load conditions, enabling the vehicle to operate in an all-electric mode.

OVCC (Off-Vehicle Charge Capable)

OVCC EVs and HVs are those that can be charged from an external power source that is not onboard the vehicle. This category includes both EVs and HEVs that can replenish their battery packs via cable-based connections to electric charging stations or standard electrical outlets [1].

OVCC vehicles are designed to utilize the electric grid for charging, thereby reducing reliance on ICEs and diminishing GHG emissions. They are equipped with charging ports that facilitate the connection to residential, commercial, or public charging infrastructure. This feature is pivotal for extending the driving range of EVs and HVs, making them more viable for longer commutes and reducing range anxiety among users.

Types:

1. **BEVs:** Full-EVs that rely solely on their battery packs for propulsion; these vehicles must be charged externally as they do not have an alternative fuel source.
2. **PHEVs:** These vehicles combine an electric motor with a traditional ICE. They can be charged externally to provide a significant electric-only drive range before switching to the ICE.

Overcurrent

Overcurrent refers to a condition where the electrical current exceeds the maximum allowable level in the circuitry of EVs and HVs. This can occur due to an abnormal or excessive demand for power, component failure, or a short circuit within the vehicle's electrical system [2].

Managing overcurrent is crucial in EVs and HVs due to the sophisticated and sensitive nature of their electrical systems, which include high-voltage batteries and complex power electronics. Overcurrent can lead to overheating, damage to components, and even safety hazards like electrical fires.

Causes:

- **Short Circuits:** The most common cause, where electrical flow deviates from the intended path, often due to damaged wires or faulty components [3].
- **Faulty Components:** Defective or malfunctioning components such as relays, switches, or ECUs can inadvertently cause overcurrent.
- **Excessive Load:** Connecting too many electrical devices or systems simultaneously can exceed the circuit's capacity.

Prevention and Protection:

- **Fuses:** Fuses are designed to blow and disconnect the circuit if current exceeds a safe level, effectively preventing further damage.
- **Circuit Breakers:** These automatically cut off electrical flow when they detect an overcurrent condition, which can be reset and reused, unlike fuses [4].
- **Current Limiters:** These devices limit the current in a circuit to a safe level, preventing overcurrent before it occurs.

Overcurrent protection is a critical aspect of designing and operating EVs and HVs. Ensuring robust protection mechanisms are in place helps in maintaining the safety, efficiency, and longevity of these advanced vehicles.

Overcurrent Protection Device

See **Fuses** and **Fusible Link**.

Overtemperature Operation

Overtemperature operation refers to conditions under which the electrical and mechanical components of EVs and HEVs operate at or above their maximum recommended temperatures. This scenario is critical because it can significantly impact the efficiency, safety, and longevity of the vehicle.

Preventative Measures:

- **Enhanced Cooling Systems:** Implementing advanced cooling technologies such as liquid cooling systems for batteries and other critical components.
- **BMS:** Integration of sophisticated BMS that monitors and regulates temperatures and implements safety protocols during critical heat events.
- **Thermal Insulation Improvements:** Improvements in thermal insulation materials and designs to help maintain optimal operational temperatures.
- **Smart Charging Strategies:** Implementing charging strategies that consider temperature conditions to minimize heat generation.

REFERENCES

[1]. Society of Automotive Engineers, "J1715 Hybrid Electric Vehicle (HEV) and Electric Vehicle (EV) Terminology," SAE Publishing, Warrendale, 2022.

[2]. S. Cash, Q. Zhou, H. Xu, O. Olatunbosun, R. Shaw and S. Davis, "New Traction Motor Sizing Strategy for an HEV/EV Based on an Overcurrent-Tolerant Prediction Model," IET Intelligent Transport Systems, 13, 168, 2019.

[3]. Z. Wu, X. Su, Y. Zhu and M. Xiao, "DC Link Capacitor Active Discharge by IGBT Weak Short Circuit," SAE Int. J. Adv. & Curr. Prac. in Mobility, vol. 1, no. 3, pp. 1177-1187, 2019, https://doi.org/10.4271/2019-01-0606.

[4]. Society of Automotive Engineers, "J553 Circuit Breakers," SAE Publishing, Warrendale, 2022.

P

Pack

A battery pack in EVs and HVs assembles individual battery cells or modules configured to store and supply electrical energy to power the vehicle's motor. It is a critical component influencing the vehicle's range, performance, and efficiency.

Battery packs typically comprise Li-ion cells chosen for their high energy density, efficiency, and longevity. The cells are assembled into modules and assembled into a larger pack with necessary electronic management systems to monitor and control performance, temperature, voltage, and current.

Key challenges in battery pack technology include improving energy density, reducing costs, enhancing safety features, and extending battery lifespans. Innovations in battery chemistry, such as solid-state batteries, and improvements in packaging and thermal management continue to advance the performance and safety of battery packs in EVs and HVs.

PEC (Power Electronics Controller)

A PEC in EVs and HVs is a critical component designed to manage and regulate the flow of electrical energy between the battery, motor, and other electrical systems. It plays a central role in converting and controlling the power supply to ensure the efficient operation of the vehicle's electric drive system.

The PEC primarily handles the conversion of DC from the battery to AC to power the electric motor and vice versa when regenerative braking occurs. This process is vital for the vehicle's performance, efficiency, and safety.

- Inverter: Converts DC from the battery to AC for the motor.
- Converter: Steps down or up voltage levels to suit different components, such as the vehicle's lighting and audio systems.
- Rectifier: Converts AC back to DC, used primarily during regenerative braking to recharge the battery.

Recent advancements in semiconductor technology, like silicon carbide (SiC) and gallium nitride (GaN), have enhanced the efficiency and performance of PECs. These materials can operate at higher temperatures and voltages, leading to more compact, reliable, and efficient controllers.

Performance Levels for EVs

Performance levels for EVs refer to the varying degrees of performance characteristics exhibited by different models of EVs. These levels encompass factors such as acceleration, top speed, range, efficiency, and handling capabilities, all contributing to the overall driving experience and suitability for different purposes.

Some parameters for performance evaluation are as follows:

- Horsepower (hp)
- Torque (Nm)
- Acceleration (0–60 mph)
- Top speed (mph)
- Range (mi)
- Battery capacity (kWh)
- Energy efficiency (MPGe)
- Regenerative braking

Permanent Magnet

A permanent magnet is a type of magnet that retains its magnetic properties over an extended period without needing an external magnetic field. In EVs, permanent magnets are commonly used in electric motors, particularly in BLDC motors and PMSMs, to generate the magnetic fields necessary for motor operation.

Permanent magnets are made from materials such as neodymium (NdFeB), samarium–cobalt (SmCo), or ferrite (ceramic magnets). These materials exhibit high magnetic strength and stability, allowing them to maintain their magnetization without significant loss over time. Permanent magnets are characterized by their ability to attract ferromagnetic materials and create a magnetic field that interacts with the motor's stator windings to produce rotational motion.

PEV (Plug-In Electric Vehicle)

A PEV is an automobile that can be recharged from any external source of electricity, such as a wall socket or dedicated charging equipment. These vehicles use one or more electric motors for propulsion. They are primarily powered by electricity stored in rechargeable battery packs, differentiating them from conventional ICE vehicles that use gasoline or diesel.

These vehicles are powered entirely by electricity and do not have a gasoline engine, fuel tank, or exhaust pipe. BEVs rely solely on electricity stored in their batteries, which must be recharged via an external electrical source.

PHEV (Plug-In Hybrid Electric Vehicle)

A PHEV is an HEV that combines a gasoline or diesel engine with an electric motor and a rechargeable battery pack. PHEVs can be plugged into an external electrical source to recharge their batteries, in addition to being able to charge them through the ICE and regenerative braking.

PHEVs have both an ICE and an electric motor. They can switch between or simultaneously use these power sources to optimize efficiency. Unlike conventional hybrids, PHEVs can drive for significant distances on electric power alone, typically ranging from 10 to 50 mi (16 to 80 km), depending on the model and battery capacity. PHEVs can be charged from the electrical grid, which allows them to use cheaper and often cleaner electric power. This charging is done via a plug in. Using electric power, PHEVs can reduce emissions of pollutants and GHGs compared to conventional gasoline or diesel vehicles, especially if the electricity comes from renewable sources. The presence of an ICE ensures that PHEVs are not limited by the range of their electric battery alone, allowing for longer trips without the need for frequent recharging.

PHEVs are particularly beneficial for drivers with short daily commutes but also need the flexibility of a conventional car for longer trips. They serve as a bridge technology between traditional ICE vehicles and full-EVs, offering a blend of lower emissions and extended range without the range anxiety associated with early EVs.

PKI (Public Key Infrastructure) Certificates

PKI certificates refer to digital certificates that establish secure and reliable communications between various components of the vehicle's ecosystem. These certificates are crucial for authenticating and encrypting data exchanged between the vehicle, charging stations, service providers, and other infrastructure elements.

PKI certificates authenticate communication between an EV or HV and the charging station, preventing unauthorized access and ensuring that charging sessions are securely billed and recorded.

In V2G communication, EVs consume power and can return energy to the grid. PKI certificates authenticate these transactions to manage and secure the energy flow, facilitating a reliable and efficient energy exchange.

Software updates over-the-air (OTA) for EVs and HVs increasingly rely on software for various functionalities, so secure OTA updates are essential. PKI certificates authenticate these updates, ensuring they are received from trusted sources without tampering.

V2X communication is for technologies that enable vehicles to communicate with other vehicles, infrastructure, and devices (collectively known as V2X); PKI certificates provide a layer of security that authenticates and encrypts the data exchanged, enhancing safety features like collision avoidance.

PMSM (Permanent Magnet Synchronous Machines)

PMSMs are a type of electric motor widely utilized in EVs and HVs due to their efficiency, compact size, and superior performance characteristics. These motors employ permanent magnets embedded in the rotor to generate a constant magnetic field, which interacts with the AC in the stator windings to produce rotational motion.

PMSMs are known for their high efficiency, particularly at variable speeds, making them ideal for the stop-and-go nature of automotive travel. Using permanent magnets allows for a more compact and lighter design compared to other types of motors, such as induction motors, which is advantageous in automotive applications where space and weight are critical considerations. These motors can deliver a high power output relative to their size, attributed to the strong magnetic field of the permanent magnets. PMSMs are capable of regenerative braking, a process that recovers kinetic energy during braking and converts it back into electrical energy to recharge the vehicle's battery, enhancing energy efficiency.

PMSMs may be primary traction motors in EVs and HVs, responsible for driving the vehicles wheels. Besides propulsion, these motors are also employed in various auxiliary systems within vehicles, such as air conditioning compressors and power steering systems, due to their precise control capabilities.

Power

For electrical systems, power is expressed in watts:

$$\text{Power}\,(\text{W}) = \text{Voltage}\,(\text{V}) \times \text{Current}\,(\text{A})$$

where

Voltage (V) is the difference in electrical potential
Current (A) is the electrical current flowing

See also **kW (Kilowatt)**.

Power Factor

Power factor, in the context of EVs and HVs, refers to the ratio of the real power (measured in watts) that is used to perform work in the vehicle's electric motor to the apparent power (measured in volt-amperes) that is drawn from the battery or power source. It measures the phase shift between line voltage and line current when the AC line is loaded with a linear load with reactive characteristics. As such, it measures how efficiently electrical power is being converted into mechanical power [1]:

$$\text{Power Factor} = \frac{\text{kW}}{\text{kVA}}$$

where
kW is the real power in kilowatts
kVA is the apparent power in kilovolt-amperes

In AC systems, which are common in EV systems, the power factor can also be expressed in terms of the cosine of the phase angle (θ) between the voltage and current waveforms:

$$\text{PF} = \cos(\theta)$$

In EVs and HVs, the power factor is crucial because it affects the efficiency and performance of the vehicle's electric drive system. A power factor close to 1 indicates that most of the power drawn from the battery is used effectively to drive the vehicle with minimal wasted energy. A lower power factor implies that a significant portion of the power is not effectively converted into practical work, often manifesting as heat or electromagnetic losses.

Power Level and Time of Charge

Time of charge refers to the duration required to recharge the battery of an EV or HV from a depleted state to total capacity. This duration depends on several factors, including the battery's capacity (measured in kilowatt-hours, kWh), the power level of the charging station (measured in kW), and the current state of the battery. Charging times can range from less than an hour using high-power DC fast chargers to several hours with standard AC home chargers. Understanding the charge time is crucial for managing travel and usage schedules for EV and HV owners [2].

Power Semiconductor Types (GaN, IGBT, MOSFET, etc.)

In EVs and HVs, power semiconductors play a critical role in efficiently converting and managing electrical energy. These components are essential for the functioning of power electronics, which handle the control and conversion of electrical power in various vehicle systems. Below are key types of power semiconductors used in EVs and HVs.

GaN semiconductors are renowned for their efficiency and high thermal conductivity. They operate at higher voltages, frequencies, and temperatures than traditional silicon-based semiconductors, which makes them ideal for high-performance applications in EVs. GaN devices are typically used in OBCs, DC–DC converters, and as part of the inverter systems that drive electric motors.

IGBTs are a type of power transistor used extensively in EV applications due to their high efficiency and ability to handle significant power loads (Figure P.1). They are especially suited for applications requiring high voltage and current, such as in EV traction inverters. IGBTs help convert DC from the battery to AC for the electric motors, ensuring efficient vehicle operation [3].

FIGURE P.1 The electronics symbol of the IGBT.

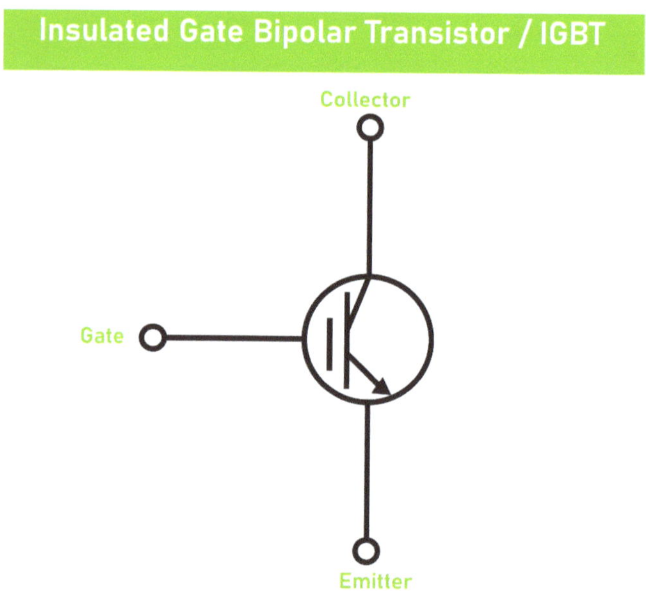

Every sketch/Shutterstock.com.

MOSFETs are another critical type of power semiconductor that offers high switching speed and lower power loss (Figure P.2). MOSFETs are commonly used in power management applications in EVs and HVs, including BMSs and smaller converters. They are preferred for lower power applications compared to IGBTs, benefiting from lower resistance and higher efficiency in specific scenarios.

FIGURE P.2 The electronics symbol of the MOSFET.

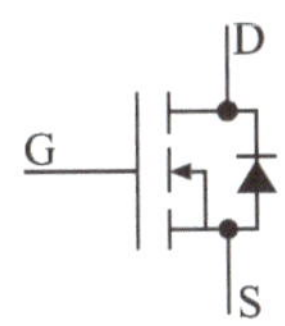

These power semiconductors are pivotal in optimizing EVs' and HVs' performance, range, and efficiency. Their ability to efficiently manage power enhances vehicle capabilities and contributes to the ongoing innovations and improvements in EV technology.

Power Split

Power split, or power splitting, is a crucial technology used in HEVs and PHEVs. It refers to distributing power from the ICE and electric motor(s) to the vehicle's wheels. This technology enables optimal utilization of both the ICE and electric propulsion systems, enhancing fuel efficiency and overall performance.

Power-Split Design Options

Power-split design options refer to the various configurations employed in the design of HEVs with power-split drivetrains. These drivetrains utilize a combination of an ICE, electric motor(s), and a planetary gear set to optimize power distribution and efficiency. Power-split designs enable seamless power blending from different sources, improving fuel efficiency, performance, and overall vehicle dynamics.

- **Series–Parallel Hybrid (Full Hybrid):**
  In this configuration, the ICE and electric motor(s) can directly drive the wheels independently or in combination.
  The planetary gear set allows for multiple operating modes, including series (electric motor driving the wheels, ICE acting as a generator), parallel (ICE directly driving the wheels), and blended modes.
- **Series Hybrid:**
  In a series hybrid design, the ICE is used solely to generate electricity to charge the battery or power the electric motor(s).
  The electric motor(s) exclusively drives the wheels, providing propulsion, while the ICE operates optimally.

- **Parallel Hybrid:**
  In a parallel hybrid configuration, the ICE and electric motor(s) are mechanically connected to drive the wheels simultaneously.
  To optimize efficiency, the planetary gear set allows for blending power from the ICE and electric motor(s) based on driving conditions.
- **Power-Split Device Variations:**
  Various arrangements of planetary gear sets, such as single or double planetary gear sets, can be employed to achieve different power-split configurations.

These configurations affect the power distribution between the ICE and electric motor(s) and determine the operational characteristics of the hybrid system.

Power-Split Mode

Power-split mode is a drivetrain configuration used primarily in HEVs, though it can also be found in some full-EVs. This system efficiently distributes power between the vehicle's ICE, electric motor, and batteries to optimize fuel efficiency and performance.

In vehicles equipped with power-split mode, the drivetrain includes a planetary gear set that acts as a CVT. The system is designed to seamlessly blend the power generated by the engine into two paths: one that directly drives the wheels and another that drives a generator. The generator can either charge the battery or power an electric motor. This seamless blending ensures the vehicle operates efficiently under various driving conditions without requiring the driver to switch modes manually.

Powertrain

The powertrain in EVs and HVs refers to the components that generate power and transmit it to the road. This system is fundamentally responsible for the vehicle's propulsion and differs significantly from traditional ICE vehicles.

Pre-Charge

The common use of pre-charge in EV context refers to charging the inverter's capacitors at a controlled rate before full application of power (closing main contractor). Failure to control charging (high inrush current) these large value capacitor (banks) causes damage to both the capacitors and possibly the supply circuitry. Thus, the power management systems engage a higher impedance pre-charge path prior to closing the main contactors. Or in some designs, the inverter manages pre-charge on its own.

PrFeB (Praseodymium Iron Boron) Permanent Magnets

PrFeB permanent magnets are a type of rare earth magnet made primarily of praseodymium, iron, and boron. These magnets are known for their exceptional magnetic strength and durability, making them highly valuable in various applications, including EV motors and wind turbine generators. PrFeB magnets are critical components in the design of high-efficiency electric motors used in HVs and EVs.

Propulsion System

The combination of the powertrain and energy supply system.

For more information, see **J1715** [1].

Protection of Persons against Electric Hazards

Protection of persons against electric hazards refers to safety measures and standards designed to safeguard users, maintenance personnel, and bystanders from electrical risks associated with EVs and HVs. These hazards include electric shock, ESS failures, and unintended energization [4, 5].

PST (Power-Split Transmission)

A PST is an advanced transmission system used primarily in HEVs and, to a lesser extent, in full-EVs. This system integrates the functions of a traditional gearbox, an electric motor, and a generator into a single unit. Its primary function is to efficiently manage power distribution between the ICE (when present), the electric motor, and the wheels.

The power-split device employs a planetary gear set that allows for a continuous variation of gear ratios without using a traditional clutch. This planetary gear set typically consists of a sun gear, planet gears, and a ring gear. In a hybrid setup, one component (often the sun gear) is driven by the ICE, another (such as the ring gear) is connected to the electric motor, and the third component (usually the planet carrier) is connected to the vehicle's drive wheels. This configuration enables the vehicle to operate in various modes: purely electric, purely ICE, or a combination of both. In full-EVs, similar setups can be used to optimize the efficiency of the electric motor at different speeds.

Public Charging Station

A public charging station, also known as an EVSE, is equipped with one or more chargers designed to recharge EVs. These stations are accessible to the general public and can vary in charging speed, connector type, and power capacity. Public charging stations are pivotal in supporting EVs' widespread adoption and convenience. The subsystems can communicate with each other via message queuing telemetry transport (MQTT) communication protocol (Figure P.3). Large-scale EV charging poses a challenge to traditional distribution networks since the extra electricity load of EV charging may overload some parts of the network [6].

See also **Duty Cycle**.

FIGURE P.3 Local MQTT publish/subscribe structure [6].

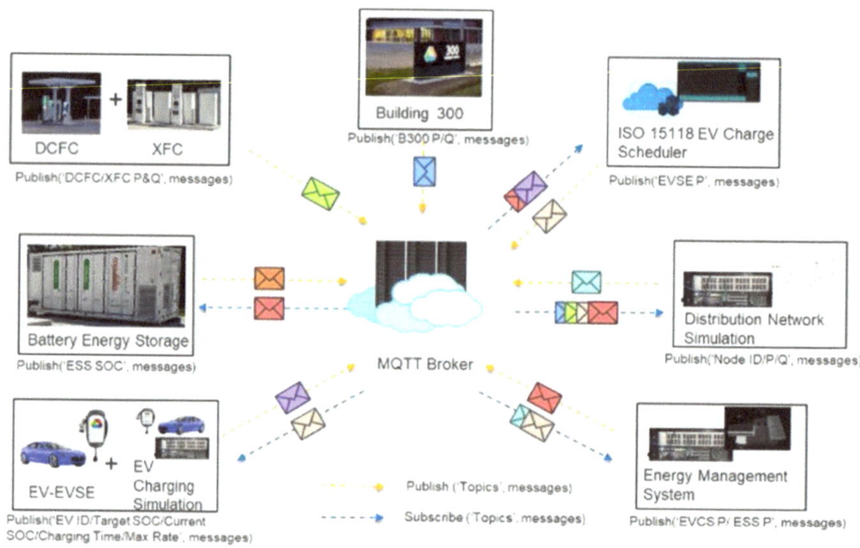

Purpose-Built Vehicle

A purpose-built vehicle is designed from the ground up with a specific function or set of tasks in mind rather than adapted from a general-use design. This type of vehicle is tailored to meet operational requirements or regulatory standards and is often utilized in commercial, industrial, or specialized mobility applications.

PWM (Pulse Width Modulation)

PWM is a method used in electronic control systems in vehicles to regulate the amount of power delivered to electronic devices. Adjusting the duty cycle—the proportion of "on" time to the regular interval or period—a PWM signal can control the amount of energy flowing to a device without analog control. This technique is extensively utilized in automotive applications to manage the speed of motors, the brightness of lights, and other electrical loads.

PWM-CP (Pulse Width Modulation–Control Pilot)

PWM-CP is a signaling method used in EV charging systems to communicate between the EVSE and the vehicle. It uses variations in the duty cycle of a pulse width modulated signal on the control pilot line to convey information about available current and control the charge process's state [7].

PZEV (Partial Zero-Emission Vehicle)

A PZEV is a classification for vehicles that produce significantly lower emissions than conventional ICEs but are not entirely emission free. This category includes full-EVs and HVs that meet specific emissions standards set by regulatory bodies such as the California Air Resources Board (CARB).

PZEVs must meet super-ultra-low-emission vehicle (SULEV) standards for exhaust emissions and have zero evaporative emissions from the fuel system.

REFERENCES

[1]. Society of Automotive Engineers, "J1715 Hybrid Electric Vehicle (HEV) and Electric Vehicle (EV) Terminology," SAE Publishing, Warrendale, 2022.

[2]. Society of Automotive Engineers, "J1772 SAE Electric Vehicle and Plug-In Hybrid Electric Vehicle Conductive Charge Coupler," SAE Publishing, Warrendale, 2024.

[3]. Z. Wu, X. Su, Y. Zhu and M. Xiao, "DC Link Capacitor Active Discharge by IGBT Weak Short Circuit," *SAE Int. J. Adv. & Curr. Prac. in Mobility*, vol. 1, no. 3, pp. 1177-1187, 2019, https://doi.org/10.4271/2019-01-0606.

[4]. Society of Automotive Engineers, "J2990 Hybrid and Electric Vehicle Safety Systems Information Report," SAE Publishing, Warrendale, 2020.

[5]. US Department of Transportation, "Lithium Battery Guide for Shippers," National Highway Traffic Safety Administration, Washington DC, 2021.

[6]. Z. Wu, N. Manne, J. Harper, B. Chen and D. Dobrzynski, "A Cloud-Based Simulation and Testing Framework for Large-Scale EV Charging Energy Management and Charging Control," *SAE Int. J. Adv. & Curr. Prac. in Mobility*, vol. 4, no. 5, pp. 1492-1500, 2022, https://doi.org/10.4271/2022-01-0169.

[7]. Society of Automotive Engineers, "J3068 Electric Vehicle Power Transfer System Using a Three-Phase Capable Coupler," SAE Publishing, Warrendale, 2024.

Q

Quick Charge

Quick charge refers to the technology and methods used to rapidly charge the batteries of EVs and PHEVs. This involves higher power levels than standard charging methods, enabling batteries to reach substantial charge levels (often 80% or more) in significantly reduced times. Quick charge systems are crucial for enhancing the practicality of EVs by minimizing downtime associated with recharging.

For more information, see **J1772** and **J3068** [1, 2].

REFERENCES

[1]. Society of Automotive Engineers, "J1772 SAE Electric Vehicle and Plug-In Hybrid Electric Vehicle Conductive Charge Coupler," SAE Publishing, Warrendale, 2024.

[2]. Society of Automotive Engineers, "J3068 Identification of Vehicles and Supply Equipment for Conductive AC Charging," SAE Publishing, Warrendale, 2023.

R

Range

Range refers to the total distance a vehicle can travel on a full battery charge or a full tank of fuel (in the case of hybrids) before requiring recharging or refueling. This is a critical metric for EVs and HEVs, as it indicates the vehicle's capability to meet various travel needs without additional energy input.

For more information, see **J1634** and **J1711** [1, 2].

Range Extender

A range extender in EVs and HVs refers to a secondary power source that increases the driving range of a BEV beyond what the primary battery can provide alone. Typically, this involves a small ICE or an additional electric generator that charges the vehicle's battery while driving.

These are small, often highly efficient petrol or diesel engines that generate electricity to recharge the vehicle's battery while driving. This type extends the vehicle's range by maintaining the battery's charge level as the vehicle is used.

Fuel cells are an attractive, range extending technology. They use hydrogen to produce electricity that charges the vehicle's battery. This technology is notable for its high energy efficiency and water production as the only by-product, presenting a greener alternative to ICE-based systems.

Though less common, some systems allow for the rapid swapping of a depleted battery for a charged one at designated stations, effectively extending the vehicle's range almost instantaneously.

Rapid Charge Paradox

The rapid charge paradox refers to the conflicting dynamics between the desire for fast charging speeds and the adverse effects these speeds can have on the battery life and efficiency of EVs and HVs.

The demand for quicker charging solutions has increased significantly. This demand has led to the development of rapid charging technologies that can charge vehicle batteries significantly in just minutes instead of hours. However, this rapid charging process can introduce a paradox.

- **Battery Degradation:** Rapid charging, especially under high currents and voltages, can lead to increased heat generation. This heat can cause accelerated degradation of the battery cells, thereby reducing the battery's overall lifespan.
- **Energy Efficiency:** While rapid charging is convenient, it is typically less energy-efficient compared to slower, steady charging methods. The faster energy transfer can lead to higher losses in the form of heat, making it a less efficient process.
- **Infrastructure Strain:** The infrastructure required to support rapid charging (such as high-power charging stations) is more complex and costly than that required for standard charging. Additionally, if many vehicles charge rapidly at the same time, it can place a significant load on the electrical grid.
- **Cost Implications:** Although rapid charging stations provide significant convenience, they are more expensive to implement and maintain. This cost can be passed on to consumers, either through higher prices for charging or in the upfront cost of the vehicle.

Rated Voltage

Rated voltage refers to the specified operating voltage at which an EV's components, particularly its battery system, are designed to operate optimally. It represents the voltage level at which the EV's electrical system is expected to function safely and efficiently under normal operating conditions.

In a typical EV, the rated voltage might range from 200 V in smaller cars to over 800 V in high-performance models. This higher voltage allows for more efficient power transmission from the battery to the motor, enhancing the vehicle's acceleration and overall performance.

Recapturing Energy

Recapturing energy, or regenerative braking or energy regeneration, is a technology employed in EVs to recover and store kinetic energy that would otherwise be lost during braking or deceleration. This process converts the vehicle's kinetic energy into electrical energy, which is then stored in the vehicle's battery for later use. Recapturing energy helps improve EVs' overall efficiency and extend their driving range.

See **Regenerative Braking**.

RECs (Renewable Energy Credits)

RECs, also known as Green Energy Certificates or Tradable Renewable Certificates, represent the environmental attributes of electricity generated from renewable energy sources. One REC is issued for every megawatt-hour (MWh) of electricity from renewable sources such as wind, solar, hydroelectric, or biomass. These credits serve as a mechanism to track and incentivize the production and consumption of renewable energy, promoting environmental sustainability and reducing carbon emissions.

Regenerative Braking

Regenerative braking is a technology used in EVs and HVs to recapture and convert kinetic energy into electrical energy during braking or deceleration. When the driver applies the brakes, the vehicle's electric motor(s) operate in reverse, acting as generators to convert the vehicle's kinetic energy back into electrical energy. This electrical energy is then stored in the vehicle's battery for later use, such as powering the electric motor during acceleration. Regenerative braking helps improve the overall efficiency of EVs by reducing energy waste and extending driving range. It also enhances braking performance by providing smoother deceleration and reducing wear on traditional friction braking systems [3, 4].

Renewable Energies

Renewable energies, also known as green energies or renewables, refer to energy sources derived from natural processes that are constantly replenished, such as sunlight, wind, rain, tides, waves, and geothermal heat. Unlike fossil fuels, which are finite and contribute to environmental degradation through GHG emissions and pollution, renewable energies offer a sustainable and environmentally friendly alternative for meeting energy needs.

Residual Charge

Residual charge refers to the remaining electrical energy stored in the battery of an EV after it has been parked or turned off. This residual charge, also known as standby or vampire charge, is present even when the vehicle is not actively being driven or charged.

The term describes the gradual loss of battery charge in EVs due to standby consumption and background systems. It emphasizes the continuous, albeit small, drain on the battery's energy reserves.

RESS (Remote Energy Storage System)

A RESS is a decentralized or off-board energy storage solution used in EVs to supplement or enhance the vehicle's onboard battery pack [2]. Unlike the primary traction battery, which is integrated into the vehicle's chassis, a RESS is a separate energy storage unit located externally. It typically consists of high-capacity batteries or other energy storage technologies housed in stationary infrastructure, such as charging stations or BSSs.

The purpose of a RESS is to provide additional energy storage capacity or charging capabilities to EVs, thereby extending their driving range, reducing charging times, or enhancing their operational flexibility. RESS solutions can take various forms. These include fast-charging stations equipped with large-scale ESSs to deliver high power or BSSs where depleted batteries are exchanged for fully charged ones.

Through RESS technology, EVs can overcome limitations associated with onboard battery capacity, such as range anxiety, long charging times, and limited access to charging infrastructure. RESS solutions offer the potential to improve EVs' usability, convenience, and scalability, supporting their widespread adoption and integration into existing transportation networks.

RESS (Remote Energy Storage System) Available Energy

RESS available energy refers to the amount of electrical energy stored and ready for use in the RESS of an EV or HEV. This metric represents the capacity of the vehicle's battery pack or energy storage unit to power the vehicle's electric motor(s) and auxiliary systems.

The RESS available energy is typically measured in kWh or megajoules (MJ). It is determined based on the battery pack's total capacity and the SOC. The SOC indicates the percentage of the battery's total capacity that is currently available for use. For example, if a battery pack has a total capacity of 60 kWh and the SOC is 80%, then the RESS available energy would be as follows:

$$SoC \times \text{Total capacity} = \text{RESS available energy}$$

$$0.8 \times 60\,\text{kWh} = 48\,\text{kWh}$$

RESS (Remote Energy Storage System) Driving

RESS driving refers to a driving mode or strategy used in EVs and HEVs to optimize the utilization of the vehicle's ESS during various driving conditions. The RESS refers to the battery pack or energy storage unit responsible for storing and supplying electrical energy to power the vehicle's electric motor(s) and auxiliary systems.

RESS driving involves intelligent management and control of the energy flow within the vehicle's electrical system, including the battery pack, motor controller, regenerative braking system, and power distribution components. RESS driving aims to maximize energy efficiency, extend driving range, and enhance overall vehicle performance by dynamically adjusting power usage based on driving demands, terrain, and environmental factors.

RESS (Rechargeable Energy Storage System) Launch

RESS Launch refers to a specialized driving mode or feature found in EVs and HEVs that optimizes power delivery from the vehicle's ESS (such as the battery pack) during acceleration, known as a launch, to enhance performance and efficiency.

RESS Launch mode is designed to leverage the maximum power output of the vehicle's ESS for quick and responsive acceleration, particularly during scenarios such as overtaking, merging onto highways, or accelerating from a standstill. The driver typically activates this driving mode or may engage automatically in certain driving conditions, optimizing powertrain performance for rapid acceleration while maintaining energy efficiency.

Reverse Power Flow

Reverse power flow refers to the capability of EVs and HVs to send stored electrical energy from their batteries back to the power grid or to other electrical devices. This process is integral to technologies such as V2G and V2H, where EVs not only draw power from the grid for charging but also have the ability to reverse the flow during certain conditions.

Reverse power flow is facilitated through an inverter that allows the DC power stored in the vehicle's battery to be converted into AC power. This AC power can then be supplied back to the power grid or used to power external devices and homes during power outages or peak electricity demand periods. The technology is seen as a potential tool for enhancing grid stability and efficiency, especially as the adoption of renewable energy sources increases.

- V2G: This technology allows EVs to interact with the power grid to return electricity during peak load times or when the vehicle is not in use. This can help stabilize the grid and provide emergency power supplies.
- V2H: This technology enables the use of the vehicle's battery to supply power to a home during outages or periods of high electricity costs, essentially turning the vehicle into a mobile power source.
- Vehicle-to-Building (V2B): Similar to V2H, this application allows vehicles to provide power to commercial buildings or facilities, which can be particularly useful in reducing energy costs and providing backup power solutions.

Reversible Machine

A device that can operate in both mechanical power output mode such as hydraulic motoring or electric motoring, and mechanical power input mode such as hydraulic pumping or electric generating. These can be hydrodynamic, hydrostatic, or electromechanical devices.

For more information, see **J1715** [5].

Roots Certificates Authority

The Roots Certificates Authority for Electric and Hybrid Vehicles is a specialized entity responsible for overseeing and managing the digital certification framework within the EV and HV industry. This authority issues and maintains digital certificates that ensure the authenticity and integrity of software updates, communications, and data exchanges between various components of EVs and HVs, such as BMSs, autonomous driving software, and connected car services.

The Roots Certificates Authority's primary role is to enhance cybersecurity measures by establishing trusted relationships between vehicle systems and external networks. By issuing digital certificates, the authority ensures that only authenticated and verified software and hardware components interact with the vehicle's sensitive systems, thereby protecting them from unauthorized access and cyber threats.

- **Issuance of Digital Certificates:** The authority provides digital certificates to vehicle manufacturers, software developers, and component suppliers. These certificates verify the entities' identities and the integrity of the content they deploy in the vehicle systems.
- **Certificate Management:** It manages the lifecycle of issued certificates, including updates, renewals, and revocations, to adapt to new cybersecurity challenges and technological advancements.
- **Standards and Compliance:** It sets and enforces standards for secure communications and data handling in EVs and HVs, aligning with global cybersecurity regulations.
- **Interoperability Assurance:** It ensures that various digital components and systems from different manufacturers can securely interact and function cohesively within the vehicle's ecosystem.

Rotor

The rotor is a critical component of the electric motor used in EVs and HVs. It is the rotating part of an electric motor positioned inside its stator.

In an EV or HEV, the rotor typically comprises a shaft with multiple attached magnets or a laminated steel core with embedded conductive coils. When electrical current flows through the motor, it interacts with the magnetic field generated by the stator, causing the rotor to spin. This rotation is transferred through the vehicle's drivetrain, propelling the vehicle forward.

See also **Electric Magnet Rotors**.

FIGURE R.1 A teardown example of the motor (especially note the rotor), with cooling jacket to improve heat characteristics [6].

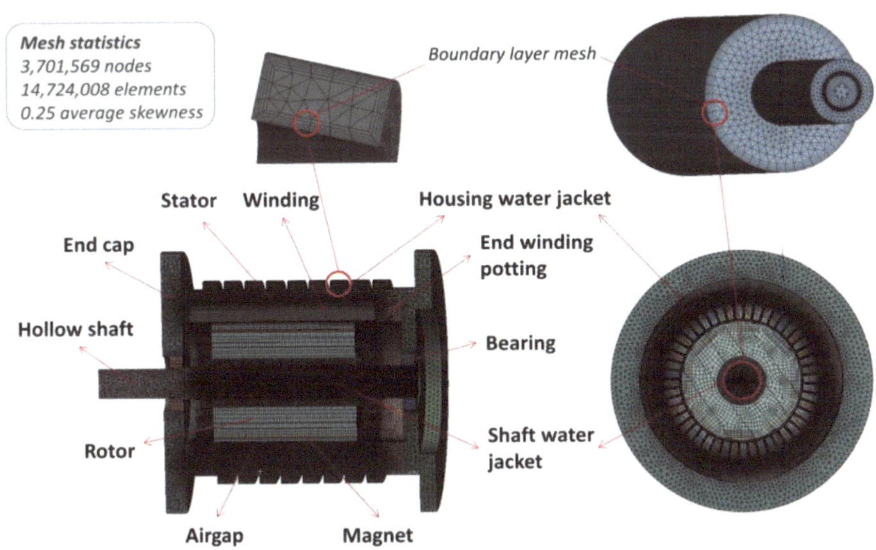

RRC (Rolling Resistance Coefficient)

The RRC is a dimensionless value that measures the resistance a tire encounters when rolling on a flat surface. It is a critical factor in the automotive industry, particularly for EVs and HVs, as it significantly influences vehicle efficiency and battery range.

EVs and HVs benefit significantly from low rolling resistance tires. Since these vehicles rely on battery power, reducing energy losses due to rolling resistance can lead to increased range and improved energy efficiency. A lower RRC means that the vehicle requires less energy to maintain a constant speed, extending the vehicle's range per charge and reducing energy consumption.

REFERENCES

[1]. Society of Automotive Engineers, "J1634 Battery Electric Vehicle Energy Consumption and Range Test Procedure," SAE Publishing, Warrendale, 2021.

[2]. Society of Automotive Engineers, "J1711 Recommended Practice for Measuring the Exhaust Emissions and Fuel Economy of Hybrid-Electric Vehicles, Including Plug-In Hybrid Vehicles," SAE Publishing, Warrendale, 2023.

[3]. S. Heydari, P. Fajri, N. Lotfi and M. Rasheduzzaman, "Maximum-Current Curve Operation of Electric Vehicles for Improved Energy Recuperation during Regenerative Braking," *SAE Int. J. Elec. Veh.*, vol. 12, no. 2, pp. 145-155, 2023, https://doi.org/10.4271/14-12-02-0007.

[4]. Y. Yao, Y. Zhao and M. Yamazaki, "Integrated Regenerative Braking System and Anti-Lock Braking System for Hybrid Electric Vehicles & Battery Electric Vehicles," *SAE Int. J. Adv. & Curr. Prac. in Mobility*, vol. 2, no. 3, pp. 1592-1601, 2020, https://doi.org/10.4271/2020-01-0846.

[5]. Society of Automotive Engineers, "J1715 Hybrid Electric Vehicle (HEV) and Electric Vehicle (EV) Terminology," SAE Publishing, Warrendale, 2022.

[6]. Q. Song, Z. Zhang and B. Ahmed, *Hollow Shaft Liquid Cooling Method for Performance Improvement of Permanent Magnet Synchronous Motors Used in Electric Vehicles*, Warrendale: SAE Publishing, 2023.

S

Safety Labeling

Safety labeling in the context of EVs and HVs refers to labels, stickers, and tags designed to provide critical safety information. These labels are applied to the vehicle's components, batteries, and other parts to warn of potential hazards, provide emergency response instructions, or indicate compliance with specific safety standards. The goal is to enhance the safety of vehicle operation, maintenance, and emergency response by clearly communicating crucial safety-related information.

For more information, see **J2894** [1], **J2929** [2], and **FMVSS 305** [3].

SC (Supercapacitor) Energy Storage

SCs, also known as ultracapacitors or electrochemical capacitors, are energy storage devices that store electrical energy through the electrostatic charge separation of ions at the interface of electrodes and an electrolyte [4]. Unlike traditional batteries, which store energy through chemical reactions, SCs store energy electrostatically, allowing for rapid charge and discharge cycles with high power density and efficiency (Figure S.1).

FIGURE S.1 Side view of the light EV [4].

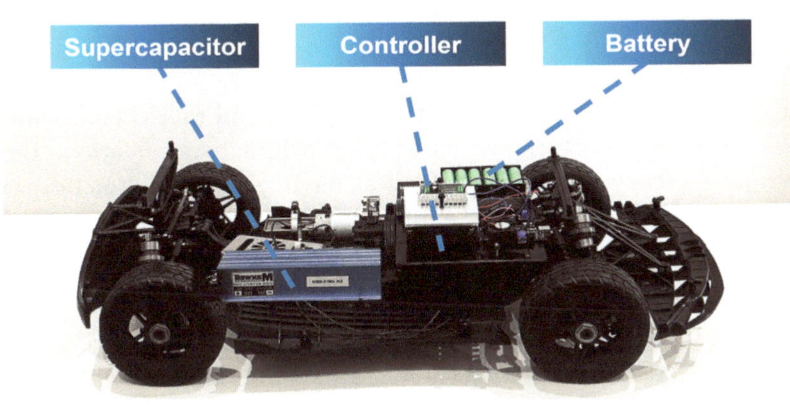

The energy storage equation for a capacitor is shown as follows:

$$E = \frac{1}{2} C V^2$$

where

E is the energy stored in the capacitor, measured in joules (J)
C is the capacitance of the capacitor, measured in farads (F)
V is the voltage across the capacitor, measured in volts (V)

In EVs, SCs can complement or replace batteries as an energy storage solution, providing high-power bursts for acceleration, regenerative braking, and other high-demand applications. SCs offer advantages such as fast charging and discharging, long cycle life, and robust performance in extreme temperatures. While they typically have lower energy density than batteries, SCs excel in applications requiring high power output and rapid energy transfer, making them well-suited for enhancing the performance and efficiency of EV drivetrains (Figure S.2).

FIGURE S.2 Battery/SC parallel configuration [4].

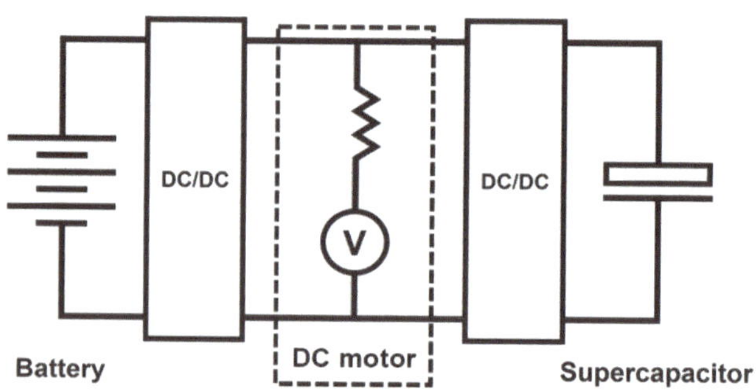

Second Use Battery

Second Use Battery refers to one that is recovered and redirected to another use when it has significantly degraded from original specification. The EV community typically defines end-of-life as having less than 80% of original capacity. For more information, see **J2854** [5], **J3071** [6], and **J2464** [7].

Secondary Batteries

Secondary or rechargeable batteries are energy storage devices used in EVs and HVs to store electrical energy that can be recharged after use. Unlike primary batteries, designed for one-time use, secondary batteries can undergo multiple charge–discharge cycles, making them ideal for sustainable energy solutions in automotive applications.

For more information, see **J2464** [7] and **J1798** [8].

Security for PEV Communication

Security for PEV communication refers to the protocols and measures implemented to safeguard the data transmitted between EVs and HEVs, charging stations, and other connected infrastructure. This aspect of automotive cybersecurity protects against unauthorized access, data breaches, and cyber-attacks that could compromise vehicle functionality, driver privacy, or network integrity (Figure S.3).

- **Encryption:** Techniques used to encode messages between vehicles and charging stations to prevent interception by unauthorized parties.
- **Authentication:** Processes to verify the identities of devices and users to ensure that communications are only established between trusted parties.
- **Data Integrity:** Ensuring the accuracy and completeness of data being exchanged, protecting against tampering or corruption during transmission.
- **Access Control:** Mechanisms that restrict access to vehicle communication systems to authorized devices and users only.
- **Anonymization:** Techniques that strip personally identifiable information from data sets to protect user privacy while allowing for system functionality and data analysis.

For more information, see **J2293** [9], **J2931** [10], **J3061** [11], and **J3101** [12].

FIGURE S.3 SAE J2293 Document Interrelationship [9].

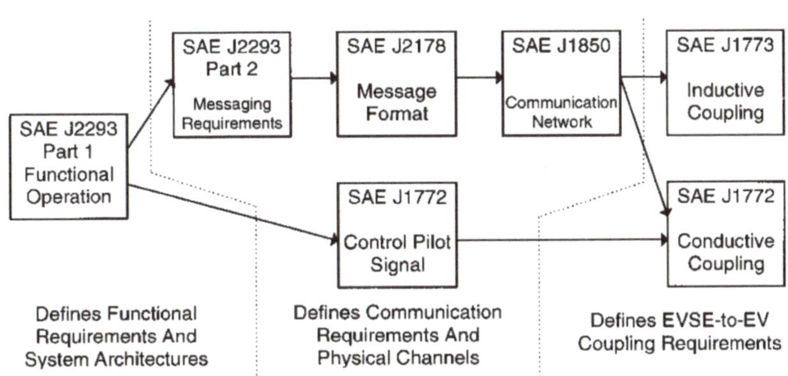

Self-Charging Hybrid

A self-charging hybrid, also known as a full hybrid, is a type of HEV that combines a conventional ICE with an electric propulsion system. This hybrid technology enables the vehicle to recharge its battery pack through regenerative braking and the ICE, without the need for external electric charging.

In self-charging HVs, the electric motor and the ICE work in tandem to optimize fuel efficiency and reduce emissions. The vehicle automatically switches between the electric motor, the ICE, or a combination of both, depending on the driving conditions. The key feature of these hybrids is their ability to generate electric power for the battery through the process of regenerative braking, which captures kinetic energy typically lost during braking and converts it into electricity.

SEP (Smart Energy Profile)

The SEP for EVs and HVs refers to a communication protocol designed to enhance the integration of these vehicles into the smart grid. This protocol facilitates efficient energy management by enabling vehicles to communicate with the electrical grid, home EMSs, and charging infrastructure.

SEP helps manage when and how an EV or HEV charges. It optimizes charging times based on utility rates, grid demand, and renewable energy availability. The protocol supports bidirectional charging (V2G), allowing vehicles to return energy to the grid during peak demand periods, which can help stabilize the grid and provide additional revenue streams for vehicle owners.

SEP (Smart Energy Profile) 2.0

SEP 2.0 is a communication protocol designed to facilitate the integration of EVs and HEVs with smart grid systems. It enables efficient energy management between vehicles and the electrical grid, supporting advanced functionalities such as DR, time-of-use (TOU) pricing, and optimized charging strategies.

SEP 2.0 is an advancement over its predecessor, offering enhanced security and interoperability among various devices and grid systems. It allows EVs and HEVs to communicate with smart grids, enabling the management of charging schedules based on grid loads and electricity pricing. This interaction helps stabilize the grid during peak loads and benefits vehicle owners from lower charging costs during off-peak hours.

Service Return and Disposal of HEV Batteries

In HVs (EVs and HEVs), service return refers to the process of returning a vehicle's battery to a service center for maintenance, repair, or replacement. This procedure is crucial for ensuring the longevity and efficiency of the battery.

Service return may involve diagnostic checks to assess the battery's SOH, capacity, and overall performance. Repairs might include module replacement, cooling system fixes, or software updates to enhance BMSs.

The disposal of HEV batteries is a critical environmental concern due to the toxic and hazardous materials they contain, such as lithium, cobalt, and nickel. Proper disposal involves several steps: First, the battery is evaluated for potential reuse in applications such as ESSs; if deemed unsuitable for reuse, it undergoes a recycling process. During recycling, valuable materials are recovered in a manner that minimizes environmental impact. This includes mechanical and chemical processes to separate and purify the components. The end goal is to safely recycle or dispose of these materials, thereby reducing the ecological footprint of EVs and HVs.

Proper service return and disposal practices are essential for the sustainability of the EV and HEV market. They not only extend the life cycle of the vehicles and their components but also mitigate the environmental risks associated with improper battery disposal. These practices are supported by regulations and guidelines in many countries, aimed at promoting responsible end-of-life management for automotive batteries.

SGIP (Smart Grid Interoperability Panel)

The SGIP is a collaborative body formed to develop and promote standards and protocols that ensure the seamless integration of EVs and HVs with the smart grid. SGIP aims to facilitate efficient communication between vehicles and the electric power grid, optimizing energy management and supporting the broader adoption of EV technologies.

The primary goal of the SGIP is to ensure that EVs and HVs can interact effectively with the smart grid. This interaction includes the ability to manage vehicle charging demands in response to grid conditions, integrate renewable energy sources, and contribute to grid stability through V2G technologies.

SHEV (Series Hybrid Electric Vehicle)

An SHEV is a type of HV where the propulsion is provided exclusively by the electric motor. The vehicle's ICE does not directly drive the vehicle's wheels; instead, it operates a generator to produce electricity. This electricity can either power the electric motor or recharge the vehicle's batteries.

In an SHEV, the ICE is typically used to generate electrical power for the electric motor, which in turn drives the vehicle's wheels. This configuration allows the ICE to operate at optimal efficiency and at a constant speed, which can improve fuel efficiency and reduce emissions. The vehicle can also be designed to utilize energy stored in the battery, allowing it to function like a pure EV over short distances.

Shore Power Code

The shore power code refers to the standards and regulations governing the connection of EVs and HEVs to an external electrical power source, typically while parked. This connection allows the vehicle to recharge its battery and operate auxiliary systems without relying on its ICE (if present) or draining its battery.

Shore power, also known as "cold ironing" in maritime contexts, has been adapted for automotive use, particularly for EVs and HVs. The code ensures safe, efficient, and standardized practices for connecting vehicles to the electrical grid or other external power sources. It covers aspects such as plug types, voltage and current requirements, safety features, and communication protocols between the vehicle and the charging infrastructure.

Single-Phase Power

Single-phase power refers to the distribution of AC electric power using a system in which all the voltages of the supply vary in unison. In EVs and HVs, single-phase power is commonly used for charging the vehicle's battery from residential and some public charging stations.

Single-phase power is a crucial component in the charging infrastructure for EVs and HVs. Most household outlets provide single-phase power, making it accessible for home charging solutions. This power setup is generally suitable for slower or Level 1 and some Level 2 charging systems, which are used when the vehicle can be parked for longer durations (e.g., overnight parking at home).

For more information, see **J2293** [9], **J1772** [13], and **J3072** [14].

Slow Charging

Slow charging refers to the method of charging EVs and HEVs at a lower power level, typically using a standard household electrical outlet. This charging method is also known as Level 1 charging in some regions.

Slow charging is primarily used by EV and HEV owners at home due to its convenience and the minimal requirements for installation. It is not typically used for quick top-ups during travel due to the lengthy time required to significantly charge the battery.

- **Voltage and Current:** Slow charging usually operates at a standard voltage of 120 V in North America or 220–240 V in Europe and other parts of the world. The current is typically around 15–20 amps.
- **Power Output:** The power output generally ranges from 1.4 to 2.4 kW, making it the least powerful but most universally accessible charging option.

- **Charging Time:** Depending on the vehicle's battery capacity and SOC, slow charging can take anywhere from 8 to 24 h to fully charge a battery. This makes it most suitable for overnight charging at home.
- **Connectors:** Commonly uses a standard three-pin plug (in the UK and similar regions) or a two-pin plug with grounding (in the US and Canada).

Smart Charging

Smart charging is an intelligent and automated approach to charging EVs that optimizes the charging process based on grid demand, electricity prices, and user preferences. Unlike traditional charging methods, which typically involve simply plugging in the vehicle and charging at a fixed rate, intelligent charging systems use advanced communication technologies and algorithms to adjust charging parameters dynamically for optimal efficiency and cost-effectiveness.

Key features of intelligent charging systems may include the following:

- **Load Management:** Smart charging systems can monitor and manage the charging load on the electrical grid to avoid overloading during peak demand periods. By scheduling charging sessions during off-peak hours or dynamically adjusting charging rates based on grid conditions, smart charging helps balance electricity demand and reduce strain on the grid infrastructure.
- **TOU Pricing:** Smart charging systems can take advantage of TOU electricity pricing, which offers lower rates during off-peak hours and higher rates during peak periods. By scheduling charging sessions to coincide with low-price periods, EV owners can minimize charging costs and use cheaper electricity rates.
- **Renewable Energy Integration:** Smart charging systems can coordinate EV charging with renewable energy generation, such as solar or wind power. By prioritizing charging during abundant renewable energy production, smart charging helps maximize the use of clean energy sources and reduce reliance on fossil fuels.
- **V2G Integration:** Some innovative charging systems enable bidirectional energy flow between EV batteries and the grid, allowing EVs to serve as mobile energy storage devices. This capability, known as V2G integration, enables EVs to store excess energy from the grid during periods of low demand and discharge it back to the grid during peak demand, providing grid stability and additional revenue opportunities for EV owners.

- **Remote Monitoring and Control:** Smart charging systems often include remote monitoring and control capabilities, allowing EV owners to monitor charging status, schedule charging sessions, and adjust charging parameters from a mobile app or web interface. This flexibility enables EV owners to optimize charging based on their preferences, energy needs, and travel schedules.

Smart Grid

A smart grid refers to an advanced electrical grid system that utilizes digital technology and automated software to monitor, control, and optimize the production, distribution, and consumption of electricity (Figure S.4). For EVs and HVs, this integration plays a crucial role in ensuring efficient energy management and reliable charging infrastructure.

The smart grid enables real-time communication between EVs and HEVs and the electricity network, allowing for efficient energy usage and management. This system supports various functionalities such as smart charging, V2G services, and enhanced grid stability.

FIGURE S.4 The smart grid has many components, from developing and routing the power to consumption.

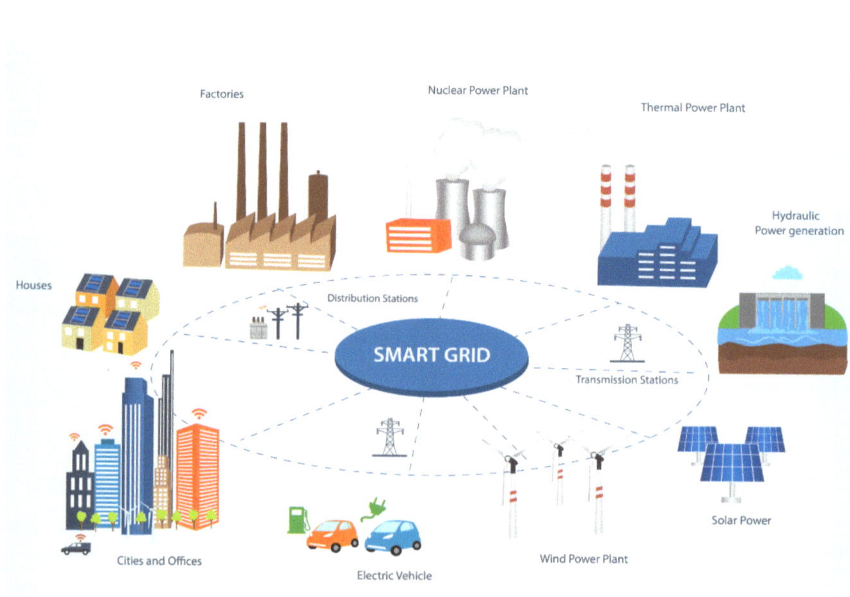

monicaodo/Shutterstock.com.

SOA (Safe Operating Area)

The SOA refers to the specified range of operational limits within which the vehicle's electrical components, such as the battery, motor, and power electronics, can function safely and effectively. An SOA violation occurs when these components are operated outside of their designated limits, potentially leading to damage, reduced efficiency, or failure.

The concept of SOA is critical in ensuring the longevity and reliability of the electrical systems in EVs and HEVs. Manufacturers define the SOA based on various parameters like temperature, voltage, current, and power. Operating outside these parameters constitutes an SOA violation, which can compromise vehicle performance and safety.

Consequences of SOA Violations:

- **Component Damage:** Persistent operation outside the SOA can result in irreversible component damage.
- **Reduced Efficiency:** SOA violations often lead to suboptimal performance, reducing the vehicle's overall efficiency.
- **Safety Risks:** Overheating, electrical shorts, and other issues arising from SOA violations can pose significant safety risks.
- **Warranty Issues:** Frequent SOA violations can void warranties or lead to increased maintenance costs.

SOC (State of Charge)

SOC is a measure used in EVs and HVs to indicate the current level of battery charge relative to its capacity, expressed as a percentage. It is crucial for managing battery performance, driving range, and overall vehicle efficiency (Figure S.5).

FIGURE S.5 Battery charging and discharging profiles [14].

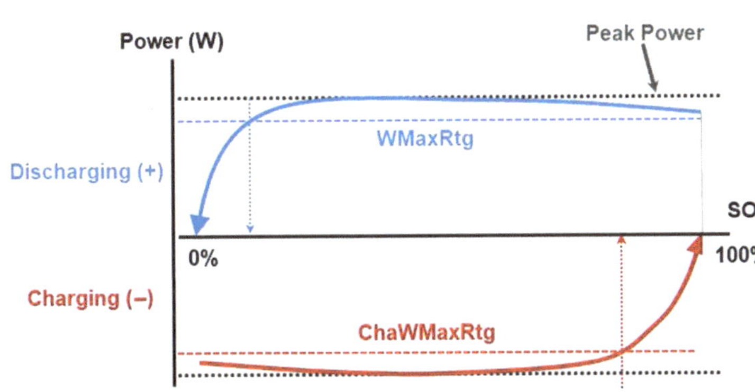

Mathematically, SOC is

$$SOC(t) = SOC(0) + \int_0^{\tau} \frac{i(t)}{C} dt$$

For more information, see **J2894** [1].

SOH (State of Health)

SOH refers to the overall condition or health status of the battery pack in an EV. It represents the battery's current capacity and performance capability relative to its original design specifications. SOH is typically expressed as a percentage, with 100% indicating that the battery is operating at its total capacity and efficiency, while lower percentages indicate degradation or loss of capacity (Figure S.6) [15].

FIGURE S.6 Illustration of request demand for BSS [15].

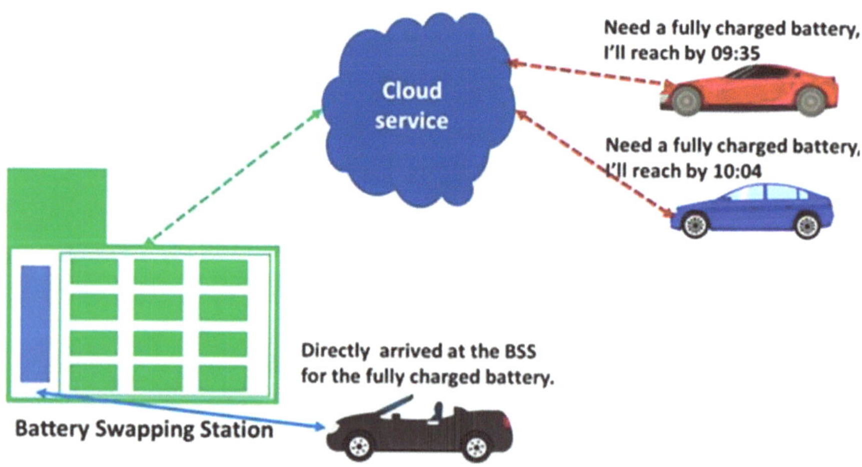

Solid-State Cell

A solid-state cell is a type of battery technology used in EV and HEV that employs solid electrolytes rather than liquid or gel-based electrolytes found in traditional Li-ion batteries. This technology offers significant improvements in energy density, safety, and longevity.

Solid-state cells represent a pivotal evolution in battery technology, addressing many of the limitations of conventional Li-ion batteries. By replacing the liquid electrolyte with a solid one, these cells can operate at higher temperatures without the risk of leakage or combustion, thereby enhancing the overall safety of the battery system.

Solid-state cells are increasingly seen as the future standard for EV battery technology due to their numerous advantages (Figure S.7). They are particularly suited for automotive applications where safety, efficiency, and high energy capacity are critical. Additionally, they have potential applications in aerospace, consumer electronics, and stationary storage systems.

FIGURE S.7 A breakdown of the solid-state cell showing the layers.

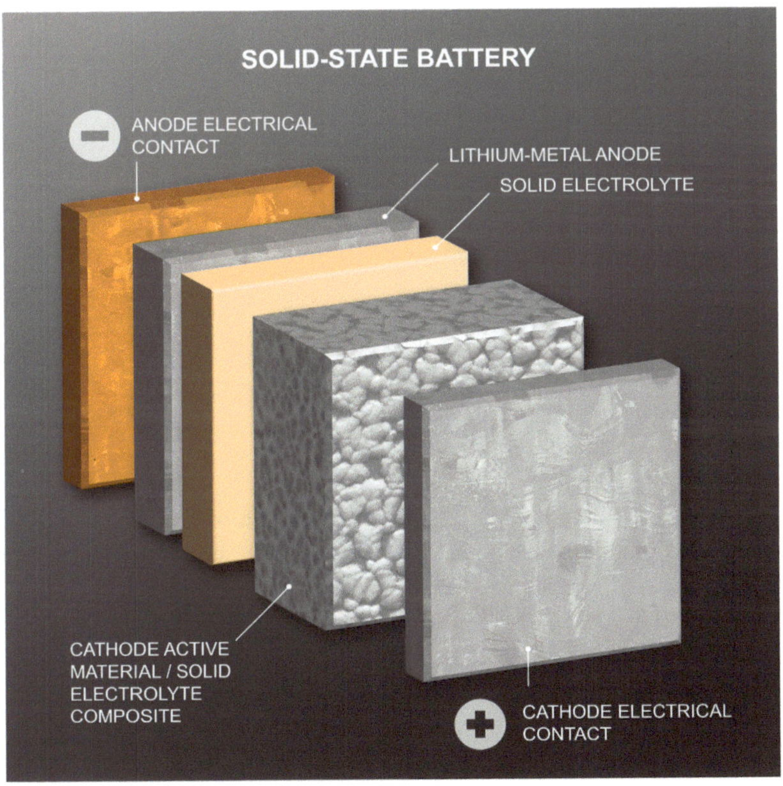

Specific Energy

Specific energy refers to the energy stored in a battery relative to its weight, typically expressed in watt-hours per kilogram (Wh/kg). This metric is crucial for assessing battery performance as it directly influences the vehicle's range, weight, and overall efficiency.

For more information, see **J1634** [16] and **J1715** [17].

Split-Phase AC Charging

Split-phase AC charging is predominantly used in North America for charging EVs and HEVs utilizing a residential electrical system offering higher power output than standard single-phase charging. This method employs a split-phase power supply to provide a 240-V AC supply, enabling faster charging rates than the standard 120-V AC outlets.

Typically, residential properties in North America are equipped with a 240-V service split into two 120-V legs. Split-phase AC charging uses this setup to double the voltage available for EV charging, significantly reducing charging time without needing commercial-grade three-phase power installations. This system is ideal for home use and compatible with most home charging equipment for EVs.

For more information, see **J1772** [13].

Split Torque Drive Transmission

Split torque drive transmission is used in EVs and HVs. It allows for the distribution of engine torque between two paths before recombining it. This technology is particularly beneficial in hybrid powertrains, where it optimizes integrating an ICE with one or more electric motors, enhancing efficiency and performance.

For more information, see **J1939** and **J3068**.

SRM (Switched Reluctance Motor)

An SRM is a type of electric motor commonly used in EVs and other applications where high efficiency, robustness, and reliability are essential. Unlike traditional electric motors that use permanent magnets or wound coils to generate magnetic fields, SRMs rely on magnetic reluctance, where the rotor is attracted to the stator's magnetic field when energized [18].

Critical features of SRMs include their simple and robust construction, which consists of a rotor with multiple salient poles and a stator with corresponding windings. Operating an SRM involves switching the current flow through the stator windings in a sequence that maximizes torque production and rotor movement [19].

SRMs offer several advantages for EVs. SRMs can achieve high efficiency over various operating conditions, making them well-suited for EV propulsion systems where energy efficiency is critical. Due to their simple construction and absence of permanent magnets, SRMs are inherently robust and less prone to demagnetization or thermal degradation, ensuring long-term reliability in demanding operating environments.

SRMs offer excellent control over torque and speed, allowing for precise modulation of power delivery to meet dynamic driving conditions and optimize energy usage in EVs.

SRMs can effectively capture and convert kinetic energy during braking into electrical energy through regenerative braking systems, enhancing overall energy efficiency and extending the driving range in EVs.

While SRMs offer significant advantages for EV applications, they also present challenges such as torque ripple and acoustic noise, which require careful design and control strategies to mitigate effectively.

Standalone Inverter

A standalone inverter in EVs and HVs is a device that converts DC from the vehicle's battery to AC, enabling the use of standard electrical appliances and equipment directly from the vehicle. This type of inverter operates independently from the vehicle's main drive system and does not require connection to an external grid.

Standalone inverters are components in EVs and HEVs where they provide energy autonomy, especially in situations where grid power is not accessible. These devices are designed to handle various power loads, depending on the specifications and the intended use within the vehicle.

For more information, see **J1772** [13] and **J551** [20].

Start/Stop

Engine is automatically turned off when power demand is zero or negative, such as during vehicle stop, and then restarted automatically. This is sometimes referenced in the industry as a micro hybrid.

For more information, see **J1715** [17].

Stator

In an EV, the stator is a stationary component of the electric motor that surrounds the rotor and generates the magnetic field necessary for motor operation. The stator typically consists of a core made of laminated steel sheets and coils of wire, known as windings, which carry electric current. When electric current is supplied to the stator windings, it creates a magnetic field that interacts with the magnetic field produced by the rotor, causing the rotor to rotate and drive the vehicle. The design and configuration of the stator play a crucial role in determining the electric motor's performance, efficiency, and power output. Efficient cooling and thermal management of the stator are essential to maintain optimal performance and prevent overheating during operation.

Swap Size, Weight, and Power

Swap size, weight, and power refers to the interchangeable battery system's dimensions, mass, and energy output characteristics. This concept is crucial for vehicles that utilize a swappable battery system, enabling easier management of vehicle range and charging time concerns.

Size refers to the physical volume or space that the battery occupies within the vehicle. This is a critical factor in design, as the battery must fit within designated compartments while leaving sufficient room for other essential components.

Weight refers to the mass of the battery, which significantly affects vehicle dynamics such as handling, acceleration, and overall efficiency. Heavier batteries might offer more energy capacity but can reduce the vehicle's performance and increase wear on components like brakes and tires.

Power is the battery's output capability, usually measured in kilowatts (kW). It determines how much energy can be delivered to the vehicle's drivetrain, directly influencing acceleration, top speed, and the ability to handle loads.

Switch-Operated Bleeder Resistors

Switch-operated bleeder resistors are components used in EVs and HVs to safely discharge electrical energy stored in the high-voltage capacitor of the vehicle's power electronics. Either passive (bleeder resistor) or active (switch-operated bleeder resistor) techniques may provide energy discharge. Passive discharge approaches are simple and reliable but inefficient and cause internal heating of inverters and storage banks. Active devices avoid inefficiency and internal heating problems but are less reliable and costlier. In either case, service personnel must verify that unacceptably high voltage is not present at the terminals of high-voltage devices before servicing [21]. These resistors are crucial for maintenance and emergency shutdown scenarios to prevent electric shock.

This switch initiates a circuit that allows the stored electrical charge in capacitors to safely dissipate through the resistor. The electrical energy is converted into heat, which is then released into the environment. These resistors' primary role is to ensure safety by reducing the system's voltage to a safe level before any service or repair work begins. This is essential in protecting maintenance personnel from electrical hazards associated with high-voltage systems typically found in EVs and HVs.

System Starter Plate

A system starter plate in EVs and HVs refers to a component of the vehicle's electrical system that is integral to the initial activation and operation of the vehicle's electric drive system. It typically functions as a connection hub or mounting platform for various electrical components essential for starting the vehicle's electric motor.

The system starter plate ensures that the electrical connections necessary for the motor's initial start-up are securely and efficiently managed. This plate may host or connect relays, fuses, and other control units crucial for the vehicle's power management and distribution. It ensures the correct activation sequence is followed and the electric systems are protected against electrical faults.

The design of the system starter plate often involves robust materials capable of withstanding the electrical loads and environmental conditions typical within a vehicle's engine compartment or chassis area, where it is usually located. The plate's design facilitates ease of maintenance and component replacement, as well as scalability for different vehicle models or power requirements.

System Voltage

System voltage in EVs and HVs refers to the electrical potential maintained by the vehicle's battery pack, which powers the electric motor and other electronic components.

The system voltage determines many aspects of an EV's or HV's performance, including the efficiency, power output, and range. Higher voltages can generally support more powerful motors. They can improve the overall efficiency of the electrical system by reducing current for a given power output, thereby minimizing energy losses due to resistance in the wiring.

There are two general categories of voltage systems. **Low-voltage systems** typically operate at voltages below 60 V. They are commonly used for smaller or auxiliary components such as electronic controls, lights, and infotainment systems. **High-voltage systems** are typically above 60 V and often between 200 and 800 V or more. These drive the main propulsion motor and handle major powertrain functions. High-voltage systems are critical for performance-oriented EVs and hybrids, as they provide the necessary power for acceleration and efficient long-range operation.

Vehicle high-voltage systems are designed with numerous safety features to protect occupants from electrical hazards. These include shielding and isolation mechanisms, fault detection systems that deactivate the voltage in an accident, and specialized maintenance tools and procedures to ensure safety during repairs. A vehicle's voltage level affects several design choices, including the type of motor used, the design of electronic controllers, and the thermal management system for the battery pack. Higher voltages often require more robust insulation and advanced cooling systems to manage the increased heat generated by higher currents.

As battery technology and electronic components evolve, system voltages may increase further to support higher power outputs and efficiency gains. This evolution is crucial for developing faster, more energy-efficient vehicles that meet the growing demands of electric transportation.

REFERENCES

[1]. Society of Automotive Engineers, "J2894 Power Quality Test Procedures for Plug-In Electric Vehicle Chargers," SAE Publishing, Warrendale, 2015.

[2]. Society of Automotive Engineers, "J2929 Safety Standard for Electric and Hybrid Vehicle Propulsion Battery Systems Utilizing Lithium-Based Rechargeable Cells," SAE Publishing, Warrendale, 2013.

[3]. National Highway Traffic Safety Administration, "FMVSS No. 305, Electric Powered Vehicles: Electrolyte Spillage and Electrical Shock Protection," US Department of Transportation, Washington, DC, 2008.

[4]. B. Noh, "Development of a Supercapacitor Hybrid Powertrain Design with Pulse-Width Modulation and Series Configuration for Light Electric Vehicles," *SAE Int. J. Elec. Veh.*, vol. 1, no. 10, pp. 79-87, 2021, https://doi.org/10.4271/14-10-01-0006.

[5]. Society of Automotive Engineers, "J2953-4 Plug-In Electrical Vehicle Charge Rate Reporting and Test Procedures," SAE Publishing, Warrendale, 2024.

[6]. Society of Automotive Engineers, "J3071 Automotive Battery Recycling Identification and Cross Contamination Prevention," SAE Publishing, Warrendale, 2016.

[7]. Society of Automotive Engineers, "J2464 (R) Electric and Hybrid Electric Vehicle Rechargeable Energy Storage System (RESS) Safety and Abuse Testing," SAE Publishing, Warrendale, 2021.

[8]. Society of Automotive Engineers, "J1798 Recommended Practice for Performance Rating of Lead Acid and Nickel Metal Hydride Electric Vehicle Battery Modules," SAE Publishing, Warrendale, 2020.

[9]. Society of Automotive Engineers, "J2293 Energy Transfer System for Electric Vehicles - Part 2: Communication Requirements and Network Architecture," SAE Publishing, Warrendale, 2014.

[10]. Society of Automotive Engineers, "J2931-4 Broadband PLC Communication for Plug-In Electric Vehicles," SAE Publishing, Warrendale, 2023.

[11]. Society of Automotive Engineers, "J3061 Cybersecurity Guidebook for Cyber-Physical Vehicle Systems," SAE Publishing, Warrendale, 2021.

[12]. Society of Automotive Engineers, "J3101 Hardware Protected Security Environment - Application Programming Interface Analysis - Information Report," SAE Publishing, Warrendale, 2024.

[13]. Society of Automotive Engineers, "J1772 SAE Electric Vehicle and Plug-In Hybrid Electric Vehicle Conductive Charge Coupler," SAE Publishing, Warrendale, 2024.

[14]. Society of Automotive Engineers, "J3072 Interconnection Requirements for Onboard, Grid Support Inverter Systems," SAE Publishing, Warrendale, 2021.

[15]. C. Gera and S. Holavanahalli, *Optimizing the Dispense Time of a Battery from the Swappable Charging Station*, Warrendale: SAE Publishing, 2024.

[16]. Society of Automotive Engineers, "J1634 Battery Electric Vehicle Energy Consumption and Range Test Procedure," SAE Publishing, Warrendale, 2021.

[17]. Society of Automotive Engineers, "J1715 Hybrid Electric Vehicle (HEV) and Electric Vehicle (EV) Terminology," SAE Publishing, Warrendale, 2022.

[18]. Neupane, N., "Comparison of Switched Reluctance Motor and Double Stator Switched Reluctance Motor," Master of Science thesis, University of New Orleans, New Orleans, 2018.

[19]. Y. Niazi, A. Gholaminejad, D. Valencia Garcia, S. Dhale and B. Nahid-Mobarakeh, *Differential Flatness-Based Control of Switched Reluctance Motors*, Warrendale: SAE Publishing, 2024.

[20]. Society of Automotive Engineers, "J551 Vehicle Electromagnetic Immunity - Electrostatic Discharge (ESD)," SAE Publishing, Warrendale, 2020.

[21]. Society of Automotive Engineers, "J2910 Recommended Practice for the Design and Test of Hybrid Electric and Electric Trucks and Buses for Electrical Safety," SAE Publishing, Warrendale, 2014.

T

TEMCH (Transmission Energy Market Clearing House)

The TEMCH is a centralized platform or entity responsible for facilitating the trading and settlement of energy transactions in an electric grid's transmission system. It serves as an intermediary between electricity generators, transmission system operators, and energy consumers, ensuring the efficient allocation of transmission capacity and the reliable operation of the grid.

In the context of EVs, the TEMCH plays a crucial role in enabling grid integration and managing the charging and discharging of EV batteries. It allows EV owners and charging station operators to participate in energy markets, offering services such as DR, ancillary services, and V2G transactions. Through the TEMCH, EVs can contribute to grid stability, balance supply and demand, and support the integration of renewable energy resources.

Temperature Monitoring

Temperature monitoring in EVs and HVs refers to the system and processes used to measure and manage the temperature of various critical components continuously. These components typically include the battery pack, electric motor, and power electronics, essential for the vehicle's performance, efficiency, and safety.

Batteries in EVs and HVs operate within a specific temperature range. Excessive heat can lead to thermal runaway, potentially causing battery degradation or failure. Temperature monitoring ensures that the battery operates within safe thermal limits.

Proper temperature management is not just about preventing overheating, it is about maintaining the vehicle's energy efficiency. By ensuring that components operate within their ideal temperature range, we can minimize energy losses due to excessive heat, thereby improving the overall vehicle range. This is a key aspect of vehicle design and maintenance that can significantly impact the performance of EVs and HVs.

See also **Thermal Management System**.

Tenant Portal

EVs rely on a rechargeable battery pack as their primary source of energy. For tenants with EVs or PHEVs, charging stations can be installed in parking areas. These stations may range from basic Level 1 chargers, suitable for overnight charging, to more rapid Level 2 chargers. Some properties might also offer DC Fast Charging options, which can significantly speed up the charging process.

Thermal Management System

A thermal management system in an EV refers to a set of components and control strategies designed to regulate the temperature of the vehicle's battery pack, electric motor, power electronics, and other critical components. Batteries operate optimally near 25°C ambient; the system control will need to heat and cool the batteries [active air cooling and heating systems (AACHS)]. These systems can be direct, air-forced through paths between the cells, or indirect; the heating and cooling are accomplished via plates that separate the battery modules [1]. The primary purpose of a thermal management system is to maintain optimal operating temperatures within these components to ensure efficient performance, prolong the component lifespan, and enhance overall vehicle safety and reliability (Figure T.1).

FIGURE T.1 Battery heat transfer mechanisms [1].

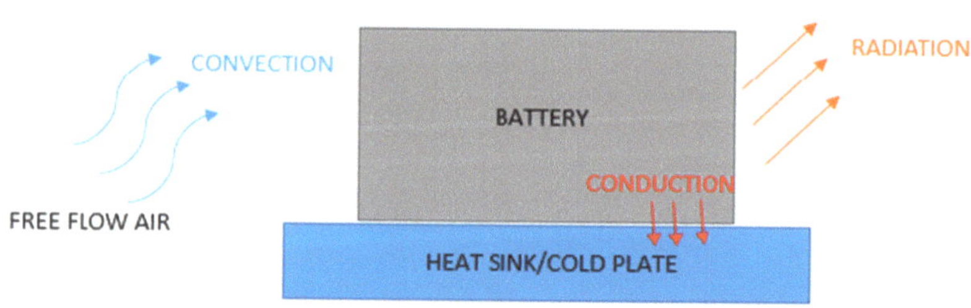

The battery cooling system uses coolant fluid, such as water or a water–glycol mixture, to absorb heat generated by the battery pack, electric motor, and power electronics. The cooling system may incorporate radiators, heat exchangers, and pumps to dissipate heat away from the components (Figure T.2).

The battery heating system provides supplemental heating to warm the battery pack and cabin during cold weather conditions, improving battery performance and passenger comfort. Heating elements or a heat pump may generate heat as needed (Figure T.3).

FIGURE T.2 Battery cooling system using a refrigerant plate [1].

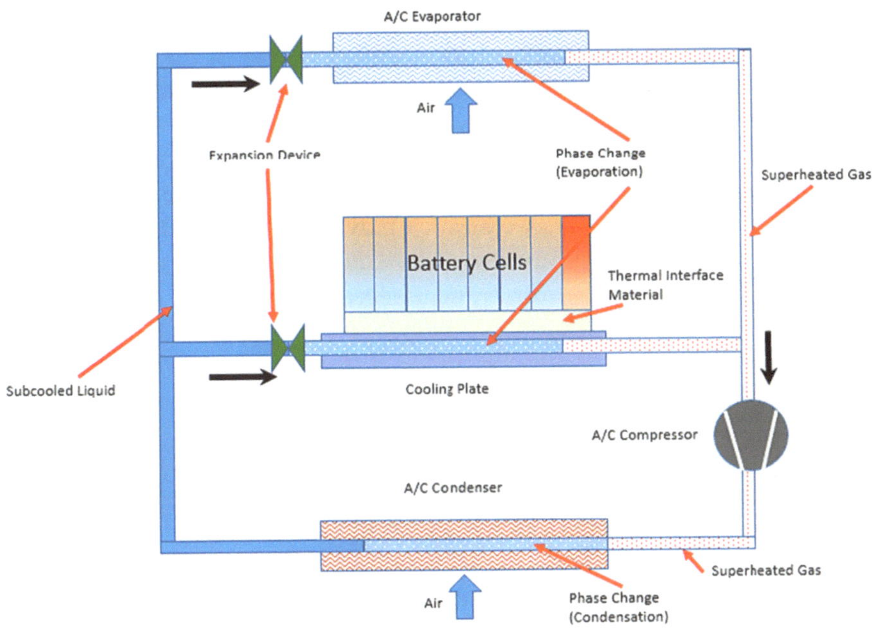

FIGURE T.3 An example of a battery heating system [1].

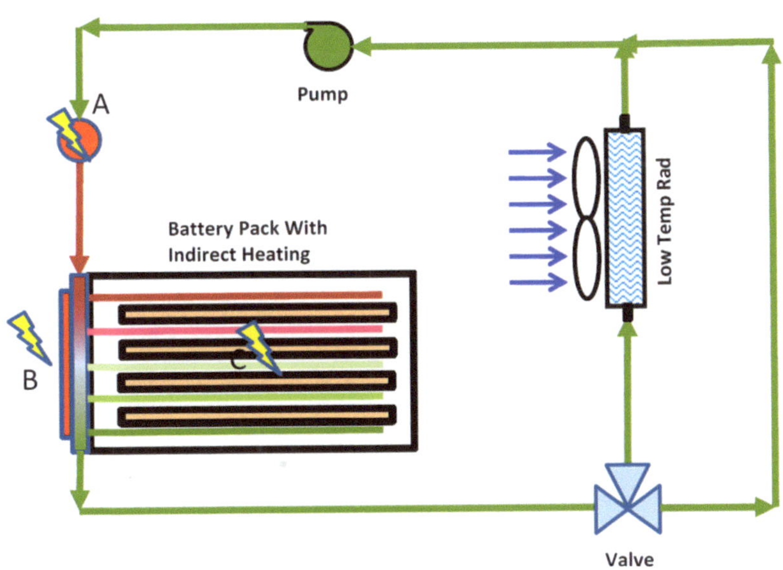

Thermal insulation minimizes heat transfer between the vehicle's interior and exterior, reducing the energy required for heating and cooling and improving overall energy efficiency. Temperature sensors monitor the temperature of critical components to ensure they remain within safe operating limits. They provide feedback to the vehicle's control system, enabling precise temperature regulation and thermal management.

The control system utilizes data from temperature sensors and other inputs to dynamically adjust cooling and heating processes as needed, optimizing thermal performance and energy efficiency. Advanced control algorithms may also incorporate predictive modeling and adaptive strategies to anticipate temperature changes and optimize system operation.

Thermal System

See **Thermal Management System.**

Three-Phase Power

Three-phase power is an electrical power transmission and distribution system that utilizes three ACs of equal frequency and amplitude, with a phase difference of 120° between phases. This configuration enables more efficient power generation, transmission, and distribution than single-phase power systems, making it suitable for high-power applications such as EV charging (Figure T.4).

FIGURE T.4 An example of the output of a three-phase power system.

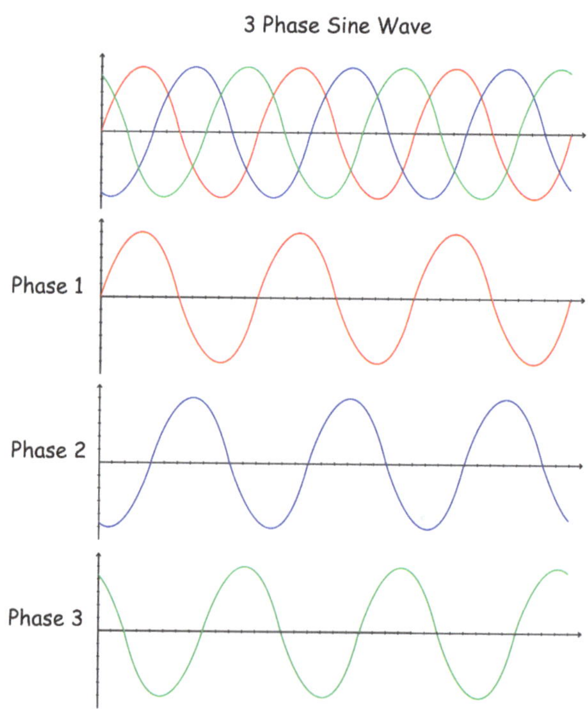

teerawat chitprung/Shutterstock.com.

EVs commonly use three-phase power for high-power AC charging, particularly in commercial and industrial settings. Three-phase AC charging stations can deliver significantly higher charging speeds than single-phase charging, allowing EVs to recharge their batteries quickly. This is particularly advantageous for fleet operators, commercial vehicle owners, and long-distance travelers who require rapid charging capabilities to minimize downtime and maximize productivity.

See also **Wye Load**.

TLS (Transport Layer Security) Session

TLS session refers to a secure communication channel established between a client and a server over a network, such as the Internet or a cellular network, using the TLS protocol. In the context of EVs and connected vehicles, TLS sessions are crucial in ensuring the confidentiality, integrity, and authenticity of data exchanged between the EV and external servers or services.

User Datagram Protocol (UDP) is a connectionless transport layer protocol that offers low latency compared to Transmission Control Protocol (TCP) but does not guarantee reliable data transfer. The UDP broadcast feature makes it useful for discovery protocols (Table T.1). In SAE J2847/6, UDP is only used by Dynamic Host Configuration Protocol (DHCP) during Internet Protocol (IP) address assignment [2].

TABLE T.1 Open systems interconnection model, a summary of layers [2].

OSI layer #	OSI layer name	Network protocol(s)	Document section(s)
7	Application layer	SAE J2847/6 application layer messages, DHCP	5.8, 7, 8, 10, 11
6	Presentation layer	JSON	5.7
5	Session layer	HTTP	5.6
4	Transport layer	TCP, UDP	5.5
3	Network layer	IPv4	5.4
2	Data link layer	IEEE 802.11	5.3
1	Physical layer	IEEE 802.11	5.2

During a TLS session, the client (such as the EV) and the server engage in a handshake process to negotiate cryptographic parameters, authenticate each other's identities, and establish a secure connection. Once the TLS session is established, data transmitted between the client and server is encrypted to prevent eavesdropping and tampering by unauthorized parties.

TLS sessions are commonly used in EVs for various purposes, including:

- **OTA Updates:** TLS sessions are utilized to securely deliver software updates and firmware patches to EVs, ensuring that the updates are transmitted securely and cannot be intercepted or modified by malicious actors.
- **Telematics and Remote Services:** EVs use TLS sessions to communicate with telematics platforms and remote service providers, enabling features such as remote vehicle monitoring, diagnostics, and control. These sessions ensure the confidentiality and integrity of sensitive vehicle data exchanged between the EV and external servers.
- **Vehicle-to-Cloud Communication:** In connected vehicle ecosystems, TLS sessions facilitate secure communication between EVs and cloud-based services, such as navigation, infotainment, and intelligent charging platforms. This ensures that data transmitted between the EV and cloud servers is protected from interception and manipulation.
- **Authentication and Authorization:** TLS sessions incorporate mechanisms for mutual authentication between the EV and the server, verifying each other's identities using digital certificates. This helps prevent unauthorized access to EV systems and ensures that only trusted servers can interact with the vehicle (Figures T.5 and T.6).

FIGURE T.5 System state diagram VA [2].

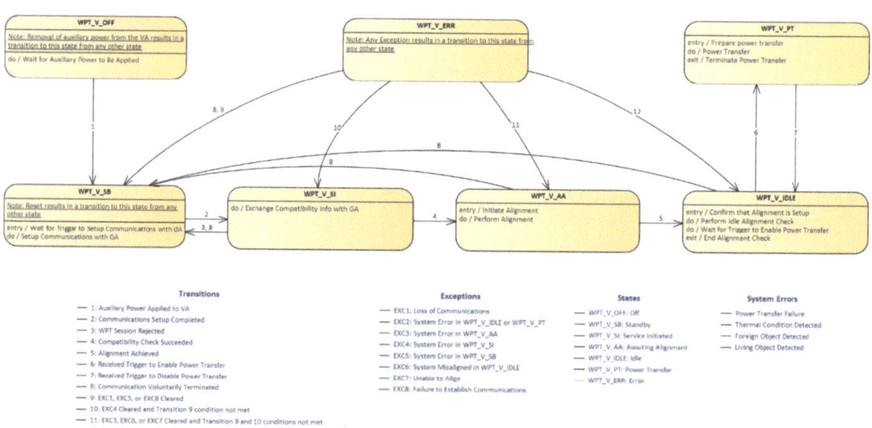

FIGURE T.6 System state diagram GA [2].

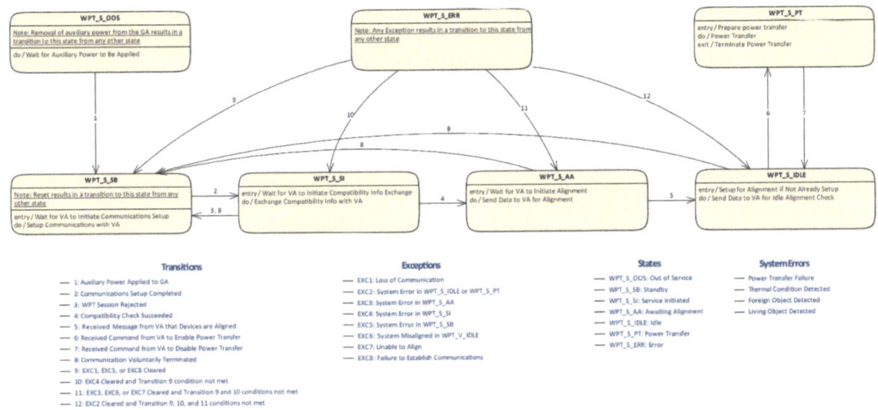

Top-Up Charging

Top-up charging refers to adding a relatively small amount of charge to the battery of an EV to increase its SOC without fully charging it. Unlike a full charge, which replenishes the battery to its maximum capacity, top-up charging only adds enough energy to meet immediate driving needs or slightly extend the range. Top-up charging is typically employed when the EV's battery is not fully depleted, and the driver desires additional range for an upcoming trip or journey. This charging strategy is convenient for EV owners who want to optimize their driving range without waiting for a complete charging cycle and is often facilitated by Level 1 or Level 2 charging stations and portable chargers. While top-up charging can provide a quick boost in range, it may not be suitable as a primary charging method for long-distance travel, as it does not fully utilize the battery's capacity.

Torque

Torque refers to the rotational force an EV's motor exerts to propel the vehicle forward. In an EV, torque is produced by the electric motor and is responsible for generating the vehicle's acceleration and overall performance. Unlike ICEs, which typically reach peak torque at specific RPM ranges, electric motors can deliver maximum torque instantaneously, providing smooth and responsive acceleration from a standstill. Torque is a crucial parameter in evaluating the performance and drivability of EVs, influencing factors such as acceleration, towing capacity, and overall driving dynamics. It is typically measured in newton-meters (Nm) or pound-feet (lb-ft) units. High torque output is often associated with powerful and dynamic driving characteristics, making it a key consideration for EV enthusiasts and performance-oriented drivers.

$$T = K_t \times I$$

where

K_t is the motor torque constant (in newton-meters per ampere)
I is the current supplied to the motor (in amperes)

The number of turns in the windings of an electric motor's stator influences the torque output. More turns generally mean a higher torque density because the magnetic field strength, which interacts with the rotor, increases with the number of turns (Figures T.7 and T.8). This means a motor can generate more torque from the same amount of electric current [3].

FIGURE T.7 Distributed windings (left) and concentrated windings (right) [3].

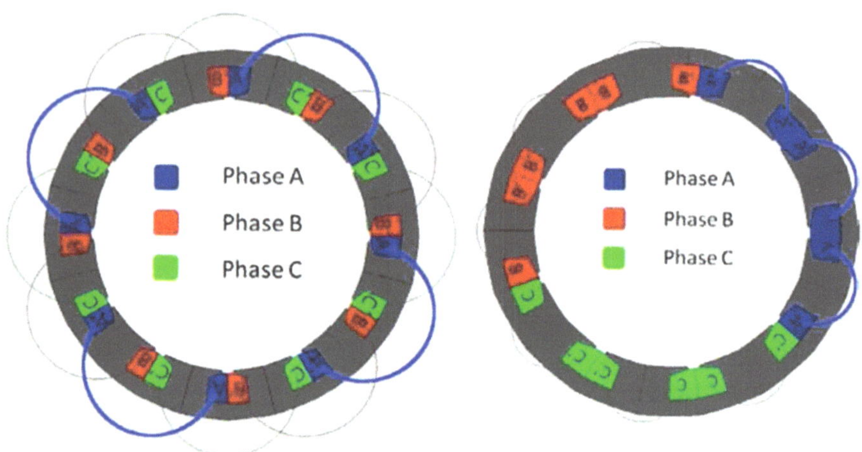

FIGURE T.8 Toroidal windings [3].

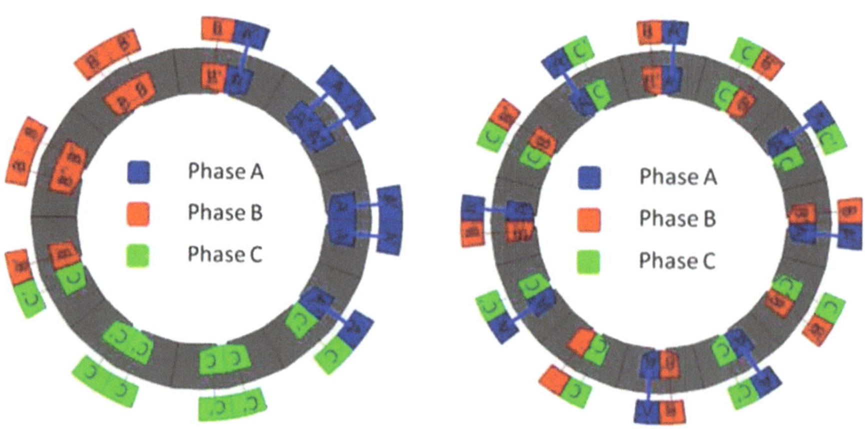

Traction Battery Pack

A traction battery pack, often referred to simply as the "battery pack," is the primary ESS used to power the electric propulsion system of an EV. It comprises individual battery cells connected in series and parallel configurations to achieve the desired voltage and capacity. The traction battery pack stores electrical energy and delivers it to the electric motor(s) to propel the vehicle.

Critical components of a traction battery pack typically include battery modules, a BMS, a thermal management system, and structural housing or casing. The battery modules contain multiple individual battery cells, which are interconnected and packaged to form a modular unit within the pack. The BMS monitors and controls various battery pack parameters, such as SOC, SOH, temperature, and voltage, to ensure safe and efficient operation (Figure T.9).

FIGURE T.9 Typical topology of a single-motor SHEV [4].

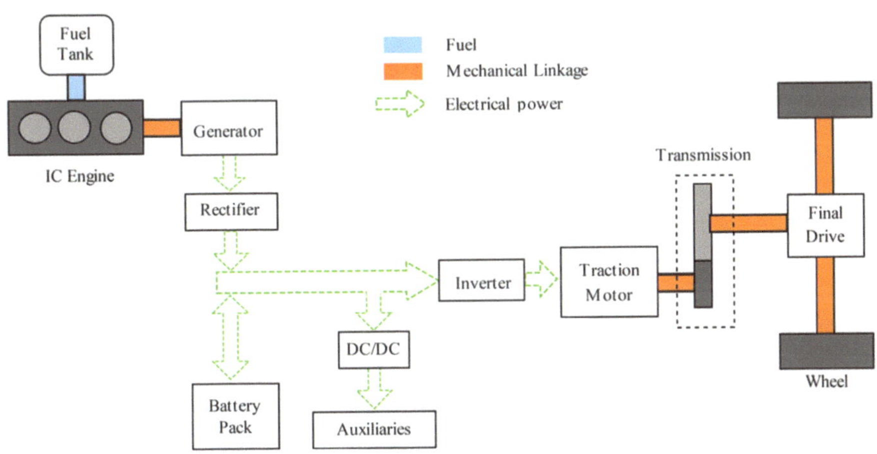

The thermal management system helps regulate the battery pack's temperature to optimize performance, efficiency, and longevity. This may involve active cooling or heating methods to maintain the battery cells within their optimal operating temperature range.

The traction battery pack is a critical EV component, directly influencing driving range, acceleration performance, and overall vehicle efficiency. Advances in battery technology, including improvements in energy density, power output, and durability, play a significant role in advancing the adoption and viability of EVs.

Traction Motor

A traction motor (or traction drive) is used in EVs and HVs to propel electricity from the vehicle's battery into mechanical energy to drive the wheels. Traction motors are critical in transportation electrification, powering the vehicle's movement and enabling efficient and environmentally friendly operation [5]. Traction motors are typically mounted directly on the vehicle's axles or integrated into the wheel hubs, delivering torque directly to the wheels to propel the car forward. They come in various designs and configurations, including AC and DC motors. They are often optimized for high efficiency, power density, and reliability to meet the performance requirements of EVs and HVs.

Transaxle

A transaxle in an EV is a combined transmission and axle assembly that integrates the transmission, differential, and drive axle into a single unit. This component is crucial in transferring power from the electric motor to the drive wheels, providing torque multiplication, speed reduction, and differential action for efficient vehicle propulsion [6].

The transaxle performs several vital functions in an EV. It receives rotational power from the electric motor and transmits it to the drive wheels through the axle shafts. Through gear ratios, the transaxle multiplies the electric motor's torque, enabling the vehicle to accelerate efficiently. The transaxle's gear mechanism reduces the motor's rotational speed to match the optimal speed range for the wheels, balancing power and speed requirements. The transaxle incorporates a differential mechanism that allows the drive wheels to rotate at different speeds during turns, ensuring smooth and stable handling.

Types of transaxles:

- **Single-Speed Transaxle:** Commonly used in many EVs, a single-speed transaxle features a fixed gear ratio optimized for typical driving conditions, offering simplicity and reliability.
- **Multispeed Transaxle:** Some high-performance or specialized EVs may feature multispeed transaxles with multiple gear ratios, which provide enhanced acceleration, efficiency, and performance across varying driving conditions.

Transmission

In the context of EVs, the term "transmission" typically refers to the mechanism responsible for transferring power from the electric motor(s) to the drive wheels. Unlike traditional ICE vehicles, most EVs do not require a conventional multispeed transmission with gears for power delivery. Instead, many EVs utilize a single-speed transmission or direct drive system, which delivers power directly from the electric motor to the wheels without shifting gear.

In an EV with a single-speed transmission, the motor's rotational speed is optimized to provide sufficient torque and power across various vehicle speeds, eliminating the need for complex gear changes. This simplifies the drivetrain design, reduces mechanical complexity, and contributes to smoother acceleration and driving experience.

While some EVs may feature multispeed transmissions for specific performance requirements, such as high-speed efficiency or towing capabilities, most modern EVs rely on single-speed transmissions or direct drive systems to deliver power efficiently and effectively to the wheels.

Transportation and Handling of Automotive RESS

Transportation and handling of automotive RESS, commonly known as EV batteries, are critical processes ensuring the safe and efficient movement of these high-energy components throughout their lifecycle, from manufacturing to disposal or recycling [7].

Transporting and handling RESS require adherence to strict safety protocols to mitigate risks associated with electrical, chemical, and physical hazards. These protocols include securing batteries to prevent movement, minimizing exposure to extreme temperatures, and protecting against mechanical damage.

Compliance with national and international regulations governing the transportation of hazardous materials is paramount. These regulations dictate RESS's packaging requirements, labeling, documentation, and transportation modes to ensure personnel safety, public health, and the environment. RESS must be packaged in durable, leak-proof, and fire-resistant enclosures during transportation and handling. Packaging materials should be selected based on their ability to withstand impact, vibration, and environmental stressors encountered during transit.

Various transportation modes, including road, rail, sea, and air, may be utilized for transporting RESS based on distance, urgency, and regulatory requirements. Each mode presents unique challenges and considerations regarding packaging, handling, and safety measures.

Personnel involved in the transportation and handling of RESS should undergo specialized training and certification programs to ensure competency in handling hazardous materials safely. Training may cover emergency response procedures, risk assessment, and proper handling techniques.

Continuous monitoring and surveillance of RESS during transportation and handling are essential to promptly detect and mitigate potential risks. Monitoring systems may include temperature sensors, leakage detectors, and Global Positioning System (GPS) tracking to ensure compliance and enhance safety.

At the end of their life cycle, RESS must be disposed of or recycled by environmental regulations. This is not just a requirement but a responsibility toward environmental sustainability. Proper disposal and recycling procedures are crucial as they minimize the ecological impact and maximize the recovery of valuable materials for reuse in new battery production. The personnel can contribute to a greener and more sustainable future by following these procedures (Figure T.10).

FIGURE T.10 SAE-recommended battery chemistry. In this figure, the blue color identifies as Li-ion-based chemistry [7].

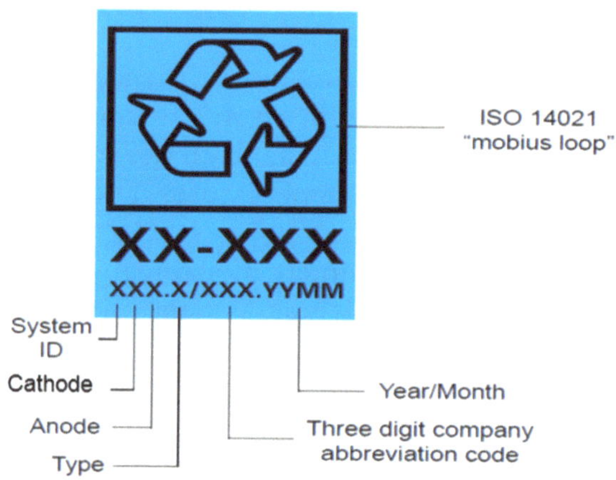

Trickle Charging

Trickle charging is a method of charging an EV battery at a low and steady rate over an extended period. It compensates for self-discharge effects and maintains the battery in an approximately fully charged state, typically for lead-acid or NiMH batteries [8]. Unlike rapid or fast charging, which delivers a high current to replenish the battery's energy quickly, trickle charging operates much slower. Typically used with Level 1 charging equipment, such as a standard household outlet, trickle charging is convenient when fast charging is unnecessary, such as overnight charging at home. While trickle charging may take longer to charge an EV battery than fully charging rapid charging methods, it is gentle on the battery. It can help extend its lifespan by minimizing heat generation and stress on the cells. Trickle charging is used as a supplementary method to maintain battery levels or as a backup option when faster charging solutions are unavailable.

Type 1 – AC

AC charging, also known as Level 1 charging, is one of the standard methods for charging EVs using a basic single-phase AC power source [9]. This charging method involves connecting the EV to a Level 1 EV charger, which is typically included with the vehicle and can be plugged into a standard residential electrical outlet.

Type 1 AC chargers operate at a standard voltage of 120 V AC. The charging current typically ranges from 8 to 16 amps, depending on the specific EV and charger capabilities. Type 1 AC charging provides a relatively slow charging rate, delivering around 2 to 5 mi of range per hour of charging, making it suitable for overnight charging or topping up during extended parking periods.

Type 1 Plug

The Type 1 plug, or the SAE J1772 connector, is a standardized charging connector used for EVs and PHEVs in North America and other regions (Figure T.11). It features a rectangular shape with five pins and is designed to facilitate the safe and reliable charging of EV batteries from AC charging stations. The Type 1 plug is compatible with Level 1 (120-V AC) and Level 2 (240-V AC) charging infrastructure, providing EV owners with convenient charging options for home charging, public charging stations, and workplace charging facilities. The connector includes safety features such as interlocking mechanisms and proximity detection to prevent accidental disconnection during charging and ensure user safety. The Type 1 plug is widely used by EV manufacturers and charging infrastructure providers, offering a reliable and standardized solution for EV charging.

For more information, see **J1772** and **J3068** [9, 10].

FIGURE T.11 An example of a Type 1 J1772 coupler [10].

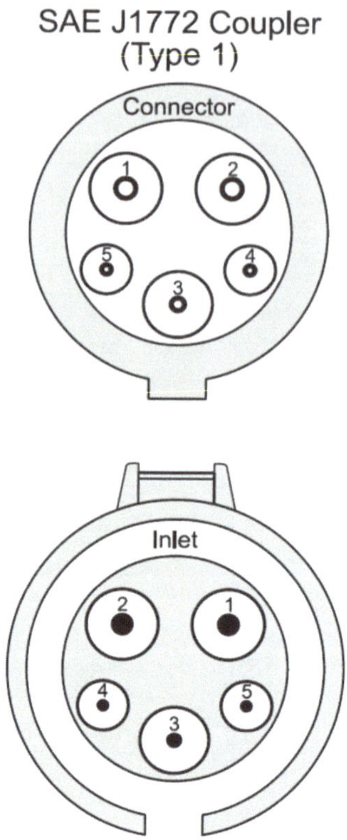

Type 2 - AC

Type 2, or AC charging, is a standard EV charging method in many regions. Type 2 AC charging supports varying charging rates, typically ranging from 3.7 to 22 kW or higher, depending on the specific charging station and the capabilities of the EV's OBC. Type 2 AC charging is compatible with various EVs, including PHEVs and BEVs. Type 2 AC charging is commonly used for home charging installations, workplace charging stations, public charging networks, and commercial charging facilities.

Type 2 Plug

The Type 2 plug is a standardized electrical connector for charging EVs in Europe and other regions (Figure T.12). It features a round, symmetrical design with seven pins arranged in a circular pattern, making it compatible with single-phase and three-phase charging systems. The Type 2 plug is widely adopted for AC charging and can deliver high power levels, making it suitable for Level 2 charging stations and home charging installations. It supports communication between the EV and the charging station, enabling functionalities like charging status monitoring, authentication, and load management. The Type 2 plug is considered a versatile and future-proof connector, supporting various charging standards and power levels, and is commonly used in EVs manufactured by European automakers.

For more information, see **J1772** and **J3068** [9, 10].

FIGURE T.12 An example of a Type 2 J3068 coupler [10].

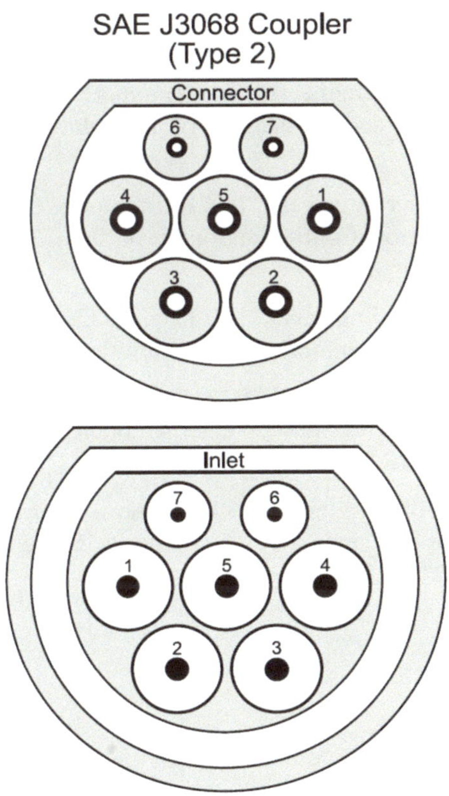

REFERENCES

[1]. Society of Automotive Engineers, "J3073 Battery Thermal Management," SAE Publishing, Warrendale, 2016.

[2]. Society of Automotive Engineers, "J2847-6 Communication for Wireless Power Transfer between Light-Duty Plug-In Electric Vehicles and Wireless EV Charging Stations," SAE Publishing, Warrendale, 2020.

[3]. Y. Tang, J. J. Paulides, E. Kazmin and E. A. Lomonova, "Investigation of Winding Topologies for Permanent Magnet In-Wheel Motors," *The International Journal for Computation and Mathematics in Electrical and Electronic Engineering*, vol. 31, no. 1, pp. 88-107, 2012.

[4]. S. Cash, Q. Zhou, H. Xu, O. Olatunbosun, R. Shaw and S. Davis, "New Traction Motor Sizing Strategy for an HEV/EV Based on an Overcurrent-Tolerant Prediction Model," *IET Intelligent Transport Systems*, 13, 168, 2019.

[5]. A. V. Brazhnikov, N. N. Dovzhenko, A. N. Minkin, O. V. Pomolotova, A. I. Litvinenko and V. A. Shilova, "Novel Type of EV Hybrid Traction Drives," *International Journal of Control and Automation*, vol. 7, no. 3, pp. 251-266, 2014.

[6]. Society of Automotive Engineers, "J645 Automotive Transmission, Transaxle, and Electrified Transmission Terminology," SAE Publishing, Warrendale, 2022.

[7]. Society of Automotive Engineers, "J3071 Automotive Battery Recycling Identification and Cross Contamination Prevention," SAE Publishing, Warrendale, 2016.

[8]. Society of Automotive Engineers, "J1715 Hybrid Electric Vehicle (HEV) and Electric Vehicle (EV) Terminology," SAE Publishing, Warrendale, 2022.

[9]. Society of Automotive Engineers, "J1772 SAE Electric Vehicle and Plug-In Hybrid Electric Vehicle Conductive Charge Coupler," SAE Publishing, Warrendale, 2024.

[10]. Society of Automotive Engineers, "J3068 Identification of Vehicles and Supply Equipment for Conductive AC Charging," SAE Publishing, Warrendale, 2023.

U

UBE (Usable Battery Energy)

The measure of RESS capacity (in kW·h) at a given point in time, indicating the remaining electric-mode driving range of an xEV.

For more information, see **J1715** [1].

UL (Underwriters Laboratories) 2202

"UL 2202 Standard for EV Charging System Equipment," is a safety standard developed by UL, specifically addressing the requirements for EV charging systems. This standard covers the electrical safety and performance criteria necessary to ensure the safe operation of EVSE, including conductive and inductive charging methods. It encompasses various aspects, such as electric shock protection, mechanical and environmental considerations, and user interface requirements. Compliance with UL 2202 is critical for manufacturers to ensure that their EV charging products meet rigorous safety standards and are fit for consumer use.

The standard applies to charging equipment in residential, commercial, and public settings. It includes but is not limited to wall-mounted units, pedestal-mounted units, and portable devices that provide AC or DC power to EVs.

UL (Underwriters Laboratories) 2231 Section 5

UL 2231 Section 5, Personnel Protection Systems for EVSE, is a critical part of the UL 2231 standard, which pertains to the safety of personnel protection systems in EVSE. This section specifically addresses the requirements for protection against electric shock and other electrical hazards when interacting with EVSE. The focus is on ensuring that the equipment incorporates adequate safety mechanisms to detect and interrupt fault conditions, thereby protecting users from potential harm.

UL 2231 Section 5's primary purpose is to define the safety criteria for personnel protection systems within EVSE. These criteria include the design and performance requirements for protective devices such as ground-fault circuit interrupters (GFCIs) and other residual current devices. The section aims to minimize the risk of electric shock to individuals using or servicing EV charging stations.

UL 2231 Section 5 applies to EVSE manufacturers and designers, providing them with guidelines to incorporate necessary safety features into their products. It ensures that all EVSEs sold and installed meet consistent safety standards, offering protection during EV charging operations. This section is particularly relevant for residential, commercial, and public EV charging station developers.

ULEZ (Ultra-Low-Emission Zone)

ULEZ aims to reduce air pollution, including nitrogen dioxide (NO_2) and particulate matter (PM), improving air quality and public health. Vehicles that do not comply with the set emission standards are either banned from entering the zone or required to pay a charge to enter.

Implementing ULEZs significantly impacts adopting and using EVs. Since EVs produce zero tailpipe emissions, they are typically exempt from any restrictions or charges imposed within ULEZs. This exemption makes EVs more attractive to consumers living in or near these zones and aligns with broader environmental goals and regulations to reduce vehicle emissions.

Ultracapacitor

See **SC (Supercapacitor) Energy Storage**.

Unshielded High-Voltage Primary Cables

Unshielded high-voltage primary cables refer to a specific category of electrical cables used in EVs to transmit high-voltage power from the battery pack to the electric motor and other high-voltage components such as the air conditioning compressor and power electronics. These cables are characterized by their lack of an electromagnetic shielding layer, which is typically employed in cabling to protect against EMI [2].

For additional information, see **J1742** and **J1113** [3, 4].

REFERENCES

[1]. Society of Automotive Engineers, "J1715 Hybrid Electric Vehicle (HEV) and Electric Vehicle (EV) Terminology," SAE Publishing, Warrendale, 2022.

[2]. IEEE Standards Association, "IEEE Std C95.-2019 IEEE Standard for Safety Levels with Respect to Human Exposure to Electric, Magnetic, and Electromagnetic Fields, 0 Hz to 300 GHz," IEEE, 2019.

[3]. Society of Automotive Engineers, "J1742 Connections for High Voltage On-Board Vehicle Electrical Wiring Harnesses - Test Methods and General Performance Requirements," Society of Automotive Engineers Publishing, Warrendale, 2022.

[4]. Society of Automotive Engineers, "J1113-1 Electromagnetic Compatibility Measurement Procedures and Limits for Components of Vehicles, Boats (up to 15 m), and Machines (Except Aircraft) (16.6 Hz to 18 GHz)," SAE Publishing, Warrendale, 2013.

V

V1G

A form of managed charging in which the charging (power level and time of charge) is adjusted to meet the prioritized needs of the EV owner and the grid [1].

V2G

A form of managed charging, where, in addition to adjusting charging behavior (V1G), a vehicle RESS can discharge to the utility grid. The charging time, charging power level, discharge time, and discharge power levels are all manageable in order to meet the prioritized needs of the EV owner and the grid [1].

V2G (Vehicle-to-Grid)

V2G is a technology that enables EVs to interact with the electrical grid, allowing them to both draw energy from the grid for charging and supply energy back to the grid when needed. In V2G systems, EV batteries serve as mobile energy storage units that can store surplus electricity during periods of low demand and feed it back into the grid during peak demand periods or when additional grid support is required. V2G technology offers several potential benefits, including grid stabilization, load balancing, and integration of renewable energy sources. By participating in V2G programs, EV owners can potentially earn revenue or incentives for providing grid services with their vehicle batteries while also contributing to the overall reliability and sustainability of the electrical grid.

V2G AC (Vehicle-to-Grid via Alternating Current)

V2G AC is a bi-directional energy transfer technology that enables EVs to not only consume electricity but also feed energy back into the electrical grid using an AC connection. In V2G AC systems, EVs equipped with bidirectional OBCs can discharge electricity from their batteries to the grid during periods of high demand or peak pricing and recharge from the grid during off-peak hours or when renewable energy generation is abundant. This technology allows EV owners to leverage their vehicles as mobile energy storage devices, providing grid stability, DR, and renewable energy integration capabilities. V2G AC systems offer potential benefits such as revenue generation for EV owners through grid services, grid balancing and stabilization, and increased utilization of renewable energy resources.

V2G DC (Vehicle-to-Grid via Direct Current)

V2G DC is a technology that enables EVs to discharge stored electrical energy back to the grid using DC charging infrastructure. In V2G systems, EV batteries act as distributed energy storage units that can provide electricity to the grid during periods of high demand or grid instability. This bidirectional flow of electricity allows EV owners to utilize their vehicle batteries for driving and grid services, such as peak shaving, frequency regulation, and DR.

In V2G DC systems, EVs equipped with bidirectional DC charging capabilities can connect to DC charging stations that support bi-directional power flow. When plugged into these stations, EVs can discharge electricity stored in their batteries back to the grid or charge from the grid, depending on grid conditions and user preferences. This flexibility enables EV owners to earn revenue by selling excess energy stored in their vehicle batteries to grid operators or utilities, contributing to grid stability and renewable energy integration.

V2G DC technology holds the potential to enhance grid reliability, support renewable energy integration, and optimize the utilization of existing infrastructure. By leveraging the energy storage capacity of EV batteries, V2G systems can help mitigate challenges associated with intermittency and variability of renewable energy sources, while also providing economic benefits to EV owners and enhancing the overall efficiency of the electric grid.

V2H (Vehicle-to-Home)

V2H technology enables EVs to serve as mobile energy storage units, supplying electricity to residential homes during power outages or peak demand periods. With V2H technology, EV batteries can function as energy reservoirs, allowing homeowners to draw power from their vehicle's battery to power essential appliances or even provide home backup electricity.

To utilize V2H technology, homeowners need a bidirectional EV charger installed. This charger must be capable of charging the EV from the grid and discharging electricity from the EV battery to the home's electrical system. Additionally, compatible EVs equipped with bidirectional charging capabilities are required to enable V2H functionality.

V2I (Vehicle-to-Infrastructure)

V2I refers to a communication framework that enables interaction between EVs and various infrastructure components within the transportation ecosystem. In V2I systems, EVs exchange data with infrastructure elements such as traffic signals, roadway sensors, charging stations, and intelligent grid systems to enhance safety, efficiency, and convenience for drivers and optimize the operation of the transportation network.

V2I technology allows EVs to receive real-time traffic information, including traffic congestion, road conditions, and construction updates, from infrastructure sensors and traffic management systems. This information can help drivers navigate more efficiently and reduce travel time.

V2I systems enable EVs to communicate with charging stations to access charging availability, pricing information, and preferred charging rates. This integration facilitates seamless charging experiences for EV drivers and supports dynamic charging management to optimize energy distribution on the grid.

V2I technology enables the exchange of safety-related information between EVs and infrastructure components, such as traffic signals and roadside sensors. This communication allows for advanced safety features, such as collision avoidance systems, intersection assistance, and pedestrian detection, to enhance overall road safety.

V2I systems facilitate bidirectional communication between EVs and the electric grid, enabling smart charging functionalities such as demand response, V2G services, and grid stabilization. V2I integration supports grid reliability, energy efficiency, and renewable energy integration by coordinating EV charging with grid conditions and renewable energy generation.

V2I data collection and analytics enable transportation agencies and urban planners to monitor traffic patterns, analyze transportation trends, and optimize infrastructure investments. By leveraging V2I insights, cities can make data-driven decisions to improve transportation infrastructure, reduce congestion, and enhance residents' mobility.

V2L (Vehicle-to-Load)

V2L technology enables EVs to serve as mobile power sources, capable of supplying electricity to external devices or systems. Unlike traditional vehicles that only consume energy, V2L-equipped EVs can bidirectionally transfer electricity, allowing them to function as portable power generators.

V2L technology allows EVs to charge from external power sources and discharge stored energy to power external devices. The V2L-enabled EVs are equipped with power outlets, typically located in the vehicle's trunk or cargo area, allowing users to connect devices directly to the vehicle's battery. The output capacity of V2L systems varies depending on the vehicle model and battery size, with some capable of delivering power ranging from a few hundred watts to several kilowatts. V2L technology enables various applications, including powering electric tools at job sites, providing emergency backup power during outages, and supporting outdoor activities such as camping or tailgating. V2L-equipped EVs can also provide grid support by feeding excess energy back to the electrical grid during peak demand periods or participating in demand response programs.

V2X (Vehicle-to-Everything)

V2X is a comprehensive communication technology that enables vehicles to communicate with various entities within their environment, including other vehicles (V2V), infrastructure (V2I), pedestrians (V2P), networks (V2N), and V2G. This connectivity allows real-time information and data exchange between vehicles and their surroundings, enhancing transportation systems' safety, efficiency, and functionality. Achieving seamless communication between vehicles and infrastructure components requires standardization and interoperability among diverse stakeholders and technology platforms.

VA (Vehicle Assembly)

A wireless power transfer assembly on the vehicle side, consisting of a coil, power electronics, control, and communication necessary to receive power from an associated GA [1].

VA (Volt-Ampere)

VA is a unit of apparent power in an electrical system commonly used in EV charging systems and equipment. Unlike real power (measured in watts), which represents the actual power consumed or delivered in an electrical system, VA represents the total power in an AC circuit, including both real and reactive power. In EV charging systems, VA is used to determine the capacity and rating of charging equipment, such as EV chargers and charging cables.

where

V is the voltage (in volts)

A is the current (in amperes) flowing through the circuit

VA is related to the power factor, which indicates the ratio of real power to apparent power in an electrical system. A low power factor indicates a higher proportion of reactive power, which may result in inefficiencies and increased losses in the system.

VAC (Volts Alternating Current) Referring to a Sinusoidal Oscillating Voltage

VAC measures electrical voltage in an AC system. In contrast to DC, which flows continuously in one direction, alternating current periodically reverses direction, creating a sinusoidal oscillating voltage waveform. VAC is a fundamental unit of measurement used to quantify the amplitude or magnitude of this oscillating voltage in electrical circuits and systems (Figure V.1).

FIGURE V.1 A comparison of DC with AC, showing the magnitude and frequency.

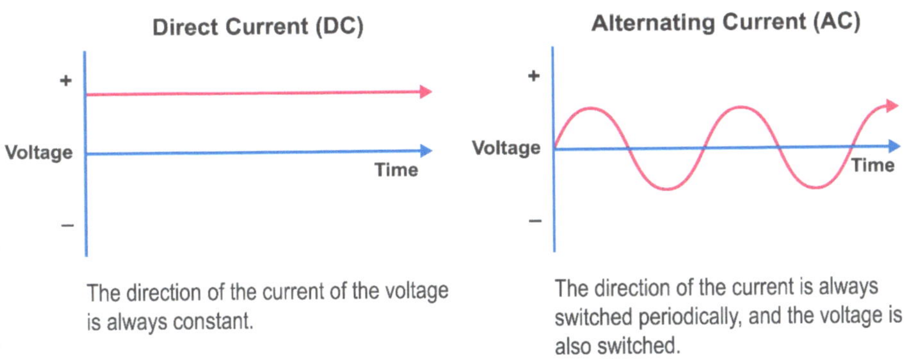

Honourr/Shutterstock.com.

Vanadium

Vanadium is a chemical element with the symbol "V" and atomic number 23. Vanadium is primarily associated with vanadium redox flow batteries (VRFBs), a rechargeable battery for energy storage applications. VRFBs utilize vanadium-based electrolyte solutions to store and release electrical energy, offering advantages such as scalability, long cycle life, and high efficiency. Vanadium is also used in certain high-strength steel alloys, which may find applications in EV manufacturing for components such as chassis, frames, and structural supports. Additionally, vanadium compounds may have potential applications in fuel cells and hydrogen production catalysts, contributing to the advancement of clean energy technologies in the automotive sector.

VCE (Vehicle Control Electronics)

VCE refers to the sophisticated electronic systems embedded within EVs that manage and control various aspects of their operation. This encompasses regulating power flow between the battery pack and the electric motor, managing battery charging and discharging processes, overseeing the vehicle's driving modes, and coordinating regenerative braking systems. VCE is pivotal in optimizing the vehicle's performance, efficiency, safety, and overall driving experience.

Vehicle Electromagnetic Immunity

Vehicle electromagnetic immunity refers to the ability of EVs to resist and operate unaffected by EMFs [2]. This characteristic is crucial in ensuring that the vehicle's electronic systems, including its propulsion system, battery management, navigation, and safety features, remain operational and reliable in environments with varying levels of EMI.

Vehicle Inlet

The vehicle inlet, often called a charging inlet or charging port, is a crucial component of connected EVs designed to facilitate the charging process. This inlet is the physical interface through which electrical power is transferred from an external power source, such as a charging station or wall outlet, to the vehicle's battery system for charging.

Vehicle Inlet (Charge)

The device on the EV into which the connector is inserted for the purpose of transferring energy and exchanging information [1].

Vehicle-Side Connection

The term vehicle-side connection refers to an EV interface where external charging equipment or infrastructure connects to the vehicle's onboard charging system. It encompasses the physical connectors, ports, and communication protocols used to establish a connection between the EV and the charging station. The vehicle-side connection allows for the transfer of electrical power from the charging infrastructure to the EV's battery pack for charging. Depending on the charging standard and the charging equipment (e.g., Level 1, Level 2, DC fast charging), the vehicle-side connection may vary in terms of connector type, voltage, current rating, and communication protocol. Common vehicle-side connections include Type 1 (SAE J1772), Type 2 (IEC 62196), CCS, CHAdeMO, and Tesla's proprietary connector. Understanding the vehicle-side connection is essential for EV owners and operators to ensure compatibility with charging infrastructure and facilitate efficient and safe charging operations.

Vehicle Towing

Vehicle towing refers to hauling or pulling a disabled or non-functioning EV using another vehicle equipped with a tow hitch or towing apparatus. Towing may be necessary when an EV experiences a mechanical breakdown, runs out of battery charge, or encounters other operational issues that prevent it from being driven under its own power.

EVs may have unique towing requirements or restrictions outlined by the manufacturer. It is essential to consult the vehicle's owner's manual or contact the manufacturer for guidance on towing procedures specific to the EV model.

Version Selection

Version selection in EVs refers to choosing between different iterations or models that EVs manufacturers offer. As the EV market continues to expand and evolve, automakers regularly introduce new versions of existing models or entirely new vehicles with upgraded features, improved performance, and enhanced technology. Version selection entails evaluating various factors such as range, battery capacity, charging capabilities, features, pricing, and overall suitability to meet the needs and preferences of the consumer.

Vertical Spectrum

Vertical spectrum refers to the range of EVs available within a specific market segment or category, characterized by varying technological sophistication, performance capabilities, and price points. The vertical spectrum encompasses various EV models, from entry-level options with basic features to premium offerings with advanced technology and luxury amenities.

The vertical spectrum includes EVs with innovative technology such as advanced battery systems, regenerative braking, autonomous driving features, and connected services. EVs across the vertical spectrum offer various performance capabilities, including varying acceleration, range, and charging speeds to cater to consumer preferences and driving needs. EVs within the vertical spectrum span a broad price range, with entry-level models offering affordability and value. In contrast, premium models command higher price points due to their advanced features and luxurious appointments. The vertical spectrum encompasses different segments of EVs, including compact cars, sedans, SUVs, crossovers, trucks, and luxury vehicles, each tailored to specific market preferences and consumer demands.

VGI (Vehicle-to-Grid Integration)

A form of managed charging in which either charging or discharging is adjusted to meet the prioritized needs of the EV owner and the power utility grid. This term is used to describe the connection of PEVs to the electrical grid under coordinated, intelligent control via digital communication to allow overall optimization of both sides of the power flow. This general term includes both V1G and V2G operations [1].

VIN (Vehicle Identification Number) Authentication, Basic Charging per Enrolled Program

VIN authentication refers to verifying the VIN of an EV to authenticate its identity and eligibility for specific services or programs. The VIN is a unique alphanumeric code assigned to each vehicle, providing a standardized way to identify individual vehicles. VIN authentication may be used in various contexts, such as EV charging networks, vehicle registration, warranty claims, and vehicle history reports, to ensure that the vehicle's identity matches the information provided and to prevent fraud or misuse.

Basic charging per enrolled program refers to a charging service or program offered to EV owners that provides access to basic charging infrastructure and services for a predetermined fee or subscription. Under this program, EV owners typically enroll or subscribe to a charging plan that offers a set level of charging services, such as access to a network of charging stations, basic charging rates, and billing features. Essential charging per enrolled program may include access to public charging stations, charging session initiation and monitoring, billing and payment processing, and customer support services. This program is designed to provide EV owners convenient and affordable access to charging infrastructure while offering flexibility and control over charging expenses.

Visibility and Accessibility of High Voltage Disconnect

The visibility and accessibility of the high-voltage disconnect refers to the design and placement of components within a connected vehicle's electrical system that allows for the safe disconnection of high-voltage power sources (Figure V.2). This critical feature ensures that maintenance personnel, emergency responders, and technicians can quickly locate and access the high-voltage disconnect point to safely disable the vehicle's high-voltage system when necessary (Figure V.3).

FIGURE V.2 Examples of high voltage warning symbols and labels [2].

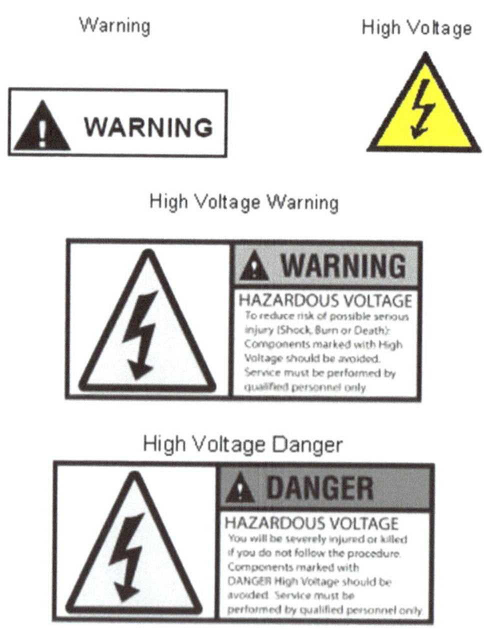

FIGURE V.3 Block diagram of the typical hybrid electric drive system [2].

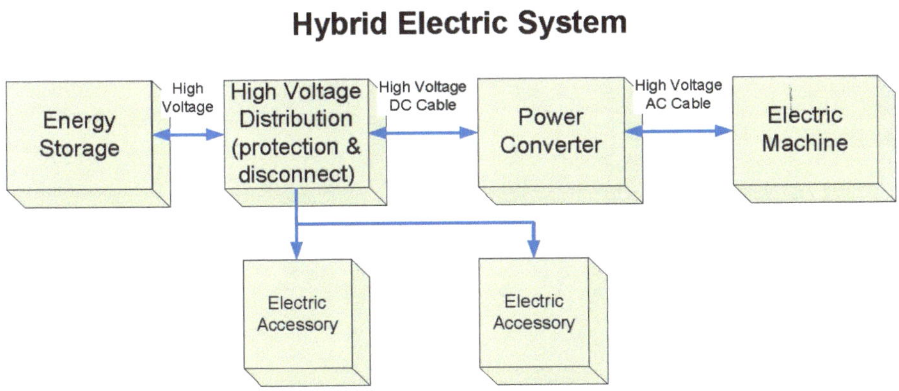

Voltage

Voltage is the potential difference between two points within the vehicle's electrical system. It is a fundamental aspect of vehicle electrification, powering various electronic components, systems, and communication interfaces essential for connectivity features.

VSC (Vehicle-Side Connection)

The connections made for the ACD system on the vehicle side of the connection. The connection is made when the infrastructure-side connection mates with the vehicle-side connection [1].

REFERENCES

1. Society of Automotive Engineers, "J1113-4 Immunity to Radiated Electromagnetic Fields - Bulk Current Injection (BCI) Method," SAE Publishing, Warrendale, PA, 2014.
2. Society of Automotive Engineers, "J2910 Recommended Practice for the Design and Test of Hybrid Electric and Electric Trucks and Buses for Electrical Safety," SAE Publishing, Warrendale, PA, 2014.

W

Wall Box

A wall box, commonly referred to as an EVSE or charging point, is a device designed to charge EVs and HVs and is installed in a fixed location, typically in a residential garage or on an exterior wall (Figure W.1).

FIGURE W.1 The wall box is mounted near where the personal vehicle is regularly parked.

alexgo.photography/Shutterstock.com.

The wall box is connected to the electrical grid and provides a safe and efficient way to charge an EV or PHEV at home. It is equipped with various safety features, such as ground fault and overcurrent protection, to ensure safe operation during charging.

A qualified electrician should install a wall box to meet local electrical codes and safety standards. This may require upgrading the home's electrical panel or adding additional wiring.

Water Exposure Tests for Insulation Breakdown

Water exposure tests for insulation breakdown are specialized testing procedures designed to assess EVs' and HVs' electrical safety and integrity when exposed to water. These tests are crucial for evaluating how well a vehicle's electrical systems, particularly its high-voltage components, can withstand moisture infiltration without experiencing insulation failure.

- **Spray and Immersion Tests:** Vehicles are subjected to water sprays (humidity) or complete immersion to mimic exposure to rain and floods. This evaluates the water-tightness of the vehicle's chassis and electrical enclosures.
- **High-Voltage Insulation Resistance Testing:** After exposure to water, high-voltage components are tested for insulation resistance using megohmmeters to ensure that insulation standards are met, indicating that the insulation has not been compromised by moisture.
- **Dielectric Testing:** This involves applying a high voltage to the electrical system to ensure that no breakdown occurs at any point in the insulation. This test helps confirm the robustness of the vehicle's electrical insulation under extreme conditions.

The electric motor operates under high voltage (Figure W.2). If the lubricant's electrical conductivity is too high, it could provide a potential path for leakage current to the housing and thereby create a shock hazard. The motor windings are made of insulated copper wire held in place by various polymeric materials. Shorting of the windings, either through deterioration of the insulation or corrosion of the copper wire, would lead to irreversible failure [1].

During water exposure tests, EV components such as wiring harnesses, connectors, and insulation materials are subjected to controlled water immersion, spray, or misting conditions [2]. The tests may vary in duration, temperature, and intensity of water exposure, depending on the specific requirements and standards applicable to the EV or component being tested.

FIGURE W.2 Examples of polymeric materials contained within electric motors [1].

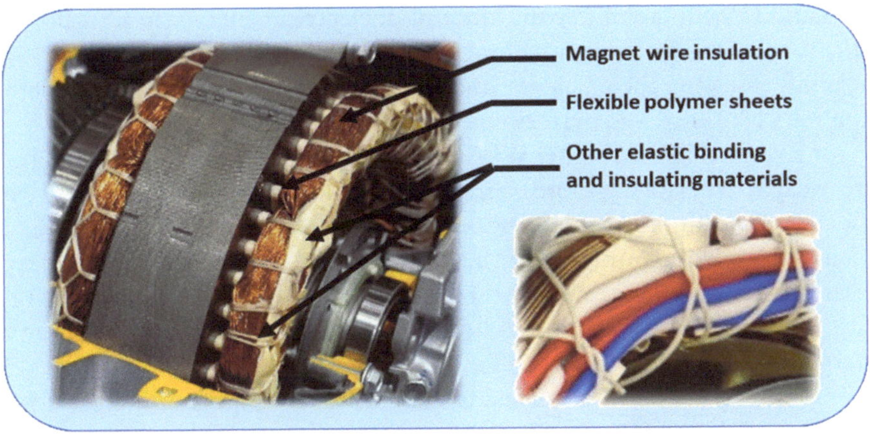

WCS (Wireless Charging System)

Conventional BMSs require a connection (wire) to each cell or parallel combination of single cells (1SxP One Serial, x Parallel.) For example, an 800 volt, Li-ion pack would require at least 250 connections (800/3.2). A wireless BMS is created by placing measurement electronics and RF networking circuitry on the same integrated circuit. These hybrid devices are then placed at each cell, cell group, or module, to greatly reduce the required BMS wiring. Various network topologies may be utilized to forward cell data from cell-to-cell and eventually to the BMS computer, or directly from each cell to the BMS computer. Other aspects of the pack such as thermal measurement may be incorporated into the network. This approach may reduce pack assembly costs and greatly reduces physical complexity. For more information, see **J1715** [3].

Wh/kg (Watt-Hours per Kilogram)

Wh/kg is a unit of specific energy commonly used to measure the energy density of batteries in EVs and HVs. It indicates the amount of energy (in watt-hours) that a battery can store or deliver per kilogram of its weight.

Wh/kg is crucial because it directly influences the vehicle's range and performance. The higher energy density (more watt-hours per kilogram) means the battery can store more energy without significantly increasing the vehicle's weight, enabling longer driving distances between charges and potentially reducing the overall vehicle weight.

Energy densities vary widely across different types of batteries. For instance, Li-ion batteries, commonly used in modern EVs, typically offer higher energy densities (around 100–265 Wh/kg) than older battery technologies like lead-acid or NiMH, which have lower energy densities.

Wh/L (Watt-Hours per Liter)

Wh/L is a unit of measurement used to express the energy density of a battery in EVs and HVs. It quantifies the amount of electrical energy (in watt-hours) that can be stored in a given volume (per liter) of the battery. This metric is crucial for assessing the efficiency and design of batteries as it directly relates to the vehicle's range and performance capabilities.

For EVs and HEVs, a higher Wh/L value indicates a more efficiently packed battery capable of storing more energy in a smaller space. This is particularly important for automotive designers and engineers striving to maximize a vehicle's range while minimizing the physical space occupied by the battery. The energy density by volume (Wh/L) complements the commonly used energy density by weight (Wh/kg), providing a comprehensive understanding of a battery's performance characteristics.

The watt-hours per liter metric is integral to developing and comparing batteries for EVs and HVs. It affects not only the potential range of the vehicle but also its design, weight distribution, and overall vehicle dynamics. Higher energy density can lead to lighter and more compact battery designs, enhancing vehicle performance and efficiency.

Wheel-End Permanent Magnet Motors

Wheel-end permanent magnet motors are also known as in-wheel motors or hub motors (PMSM) [4]. Wheel-end motors are integrated directly into an EV's wheels. Unlike traditional EV designs, where motors are housed within the vehicle's chassis and connected to the wheels via a drivetrain, wheel-end motors are situated at the wheel hubs, providing a direct drive mechanism. Integrating wheel-end permanent magnet motors directly into the wheel hubs eliminates the need for complex drivetrain components such as axles, transmissions, and differentials. This integration simplifies the vehicle's mechanical design and reduces overall weight and complexity [5].

One key benefit of wheel-end motors is their direct drive configuration, which eliminates mechanical losses associated with traditional drivetrains, thereby improving efficiency and responsiveness. This configuration also enables precise torque control and regenerative braking, enhancing vehicle dynamics and performance. Furthermore, placing the motors at the wheel hubs frees space within the vehicle's chassis, allowing for optimized interior layout and packaging. This can increase passenger comfort, cargo space, and overall vehicle versatility.

Wheel-end motors offer the potential for independent control of each wheel, enabling advanced vehicle dynamics features such as torque vectoring, traction control, and stability control. This precise control enhances vehicle stability, handling, and safety in various driving conditions.

Like traditional electric motors, wheel-end permanent magnet motors can function as generators during deceleration, converting kinetic energy into electrical energy and storing it in the vehicle's battery pack.

See **Regenerative Braking.**

Wireless BMS (Battery Management System)

See **DRFPI (Direct Radiofrequency Power Injection).**

Withstand Voltage AC

Withstand voltage AC refers to the maximum voltage level that an electrical component, such as a cable, connector, or insulation material, can withstand without experiencing electrical breakdown or failure when subjected to an AC electrical system. It is a crucial parameter in EV components, as it ensures the safety and reliability of the electrical system under normal operating conditions and potential electrical faults [6]. Withstand voltage AC is typically measured in volts (V) and represents the insulation capability of the component to withstand transient voltage surges and continuous electrical stress without compromising its integrity or performance [7]. Components with higher withstand voltage AC ratings can operate safely in high-voltage EV systems and are essential for maintaining electrical insulation and preventing electrical hazards such as short circuits or electrical arcing.

For more information, see **J2501** [8], **J2344** [9], **J1654** [10], and **J2183** [11].

WLTP (Worldwide Harmonized Light Vehicle Test Procedure)

The WLTP is an internationally recognized standard for testing the fuel consumption, CO_2 emissions, and electric range of light-duty vehicles, including EVs. Developed by the United Nations Economic Commission for Europe (UNECE), WLTP aims to provide more accurate and reliable measurements of vehicle performance and emissions by simulating real-world driving conditions more closely than previous testing procedures.

WLTP incorporates a series of laboratory-based test cycles that replicate various driving scenarios, including urban, suburban, and highway driving, as well as different vehicle speeds and accelerations. The test cycles are designed to reflect typical driving behavior and environmental conditions, resulting in more representative measurements of vehicle fuel efficiency, emissions, and electric range.

For EVs, WLTP testing includes specific procedures for measuring electric energy consumption and determining the electric range based on battery capacity and driving conditions. By providing standardized testing protocols, WLTP enables consumers to compare the performance and efficiency of different vehicle models more accurately, facilitating informed purchasing decisions and promoting the adoption of low-emission and EVs worldwide.

Working Voltage

The working voltage refers to the range of electrical potential or voltage at which an EV or its components operate safely and effectively. It is typically specified in volts (V). It represents the voltage levels within which the electrical systems, components, and devices of the EV function optimally without risk of damage or malfunction. The working voltage encompasses the voltage range of the vehicle's battery pack, electrical propulsion system, auxiliary systems, and onboard electronics. Understanding the working voltage is essential for ensuring compatibility, safety, and proper functioning of electrical components and systems in EVs. It also guides the design, operation, and maintenance of EVs to optimize performance and reliability while minimizing risks associated with electrical operation.

Workplace Charging Scheme

The Workplace Charging Scheme is a UK government initiative that supports the installation of EV charging infrastructure in workplaces. Under the scheme, eligible businesses and organizations can receive financial assistance toward the upfront costs of purchasing and installing EV charging stations for their employees and visitors. The Workplace Charging Scheme offers a grant of up to 75% of the purchase and installation costs, capped at a maximum of £350 per socket, with a maximum of 40 sockets per business location. By incentivizing the deployment of EV charging infrastructure in workplaces, the scheme aims to encourage the adoption of EVs and facilitate convenient and accessible charging options for EV owners while at work. The Workplace Charging Scheme is administered by the Office for Zero Emission Vehicles (OZEV). It is part of the UK government's broader efforts to promote the transition to EVs and reduce GHG emissions from transportation.

WP.29 (World Forum for Harmonization of Vehicle Regulations)

WP.29, also known as the World Forum for Harmonization of Vehicle Regulations, is an international regulatory forum established under the auspices of the UNECE. It serves as a platform for member countries to develop and harmonize technical regulations and standards related to vehicle safety, environmental performance, and vehicle construction.

WP.29's primary objective is to facilitate the mutual recognition of vehicle regulations among member countries, thereby promoting global harmonization and facilitating international trade in vehicles and automotive components. By establishing common technical requirements and testing procedures, WP.29 aims to enhance vehicle safety, reduce barriers to market access, and promote environmental sustainability.

WP.29 develops and maintains a set of international vehicle regulations known as the UN Regulations, which cover various aspects of vehicle design, construction, and performance, including safety features, emissions standards, and crashworthiness requirements. These regulations are adopted by member countries through a mutual recognition process, allowing vehicles certified to comply with UN Regulations to be sold and operated in multiple countries without the need for additional testing or certification.

For EVs, WP.29 plays a crucial role in developing regulations and standards specific to EV safety, performance, and interoperability. This includes regulations related to EV battery safety, electric drivetrain components, charging infrastructure compatibility, and V2G communication systems.

WPT (Wireless Power Transfer)

WPT for EVs typically involves inductive charging, transferring power from a transmitting coil to a receiving coil without physical contact. The basic equation governing the operation of a WPT system is derived from Faraday's law of electromagnetic induction and can be represented as follows:

$$V = N \times \left(\frac{d\Phi}{d\tau} \right)$$

where

V is the induced voltage in the receiving coil

N is the number of turns in the receiving coil

$\left(\frac{d\Phi}{d\tau} \right)$ is the rate of change in magnetic flux through the coil

For WPT in the context of EV charging, the efficiency of power transfer is crucial, and it can be expressed as follows:

$$\eta = \left(P_{out} / P_{in} \right) \times 100\%$$

where

η is the efficiency of the WPT

P_{out} is the power received by the receiving coil (and utilized to charge the vehicle's battery)

P_{in} is the power supplied to the transmitting coil

The actual power transferred wirelessly can also be related to the coupling coefficient between the two coils (k), the quality factor of the coils (Q), and the load connected to the system. The resonant frequency of the system, which is crucial for optimizing the efficiency of power transfer, is given as follows:

$$f_{res} = \frac{1}{\left(2\pi \times \sqrt{(LC)}\right)}$$

where
f_{res} is the resonant frequency
L is the inductance of the coils
C is the capacitance added to the coils to achieve resonance

These equations are fundamental in designing a practical WPT system for EVs. The system must achieve optimal power transfer efficiency and ensure the vehicle's battery can be charged effectively and safely without a physical connection.

See also **DRFPI (Direct Radiofrequency Power Injection)**.

WTW (Well-to-Wheel) Efficiency

WTW efficiency is a comprehensive measure used to assess the overall energy efficiency of EVs and HVs from the initial energy source to the vehicle's wheels (Figures W.3 and W.4). This metric includes all stages of energy production, processing, distribution, and final use in the vehicle, providing a complete picture of the propulsion system's environmental impact and energy efficiency.

FIGURE W.3 Generalized energy flow diagram for tank-to-wheel energy efficiency calculation of a BEV [12].

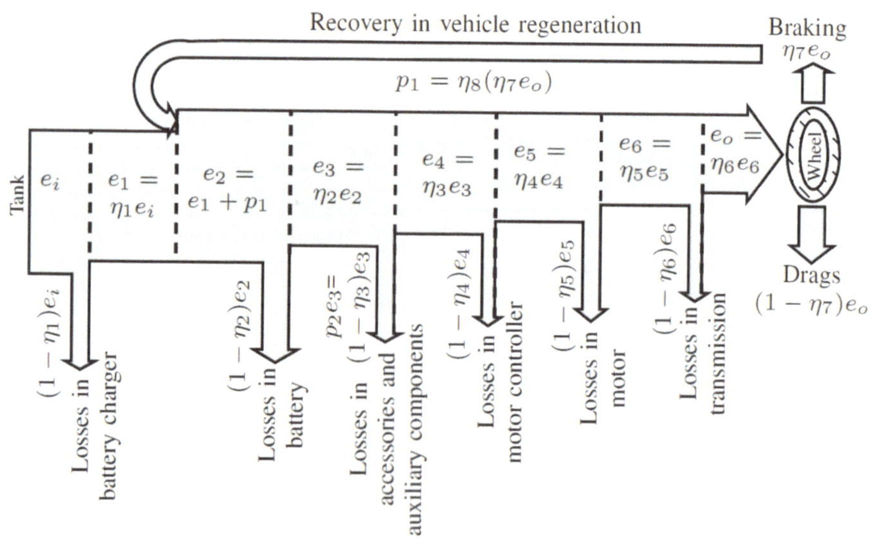

FIGURE W.4 Energy flow diagrams for WTW energy efficiency of a four-wheel independent drive BEV (upper chart), a single-motor drive BEV (middle chart), and an ICE vehicle (lower chart) [12].

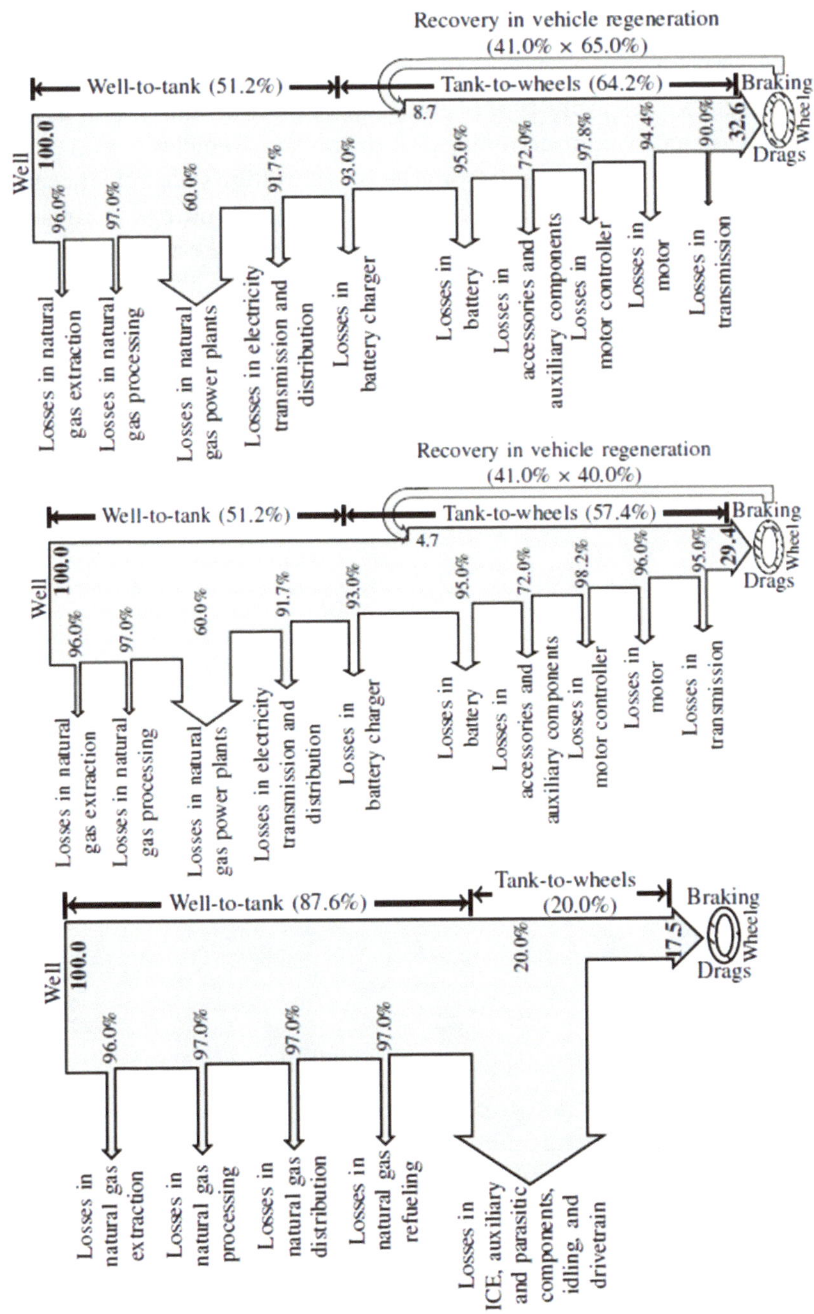

WTW efficiency is critical for evaluating EVs' and HVs' true environmental and energy efficiency benefits. It helps understand the total emissions of GHGs and pollutants over the vehicle's operational lifecycle, thus providing a basis for comparing different vehicle technologies and energy policies.

Wye Load

A wye load refers to a specific type of electrical load connection found in three-phase electric motors commonly used in propulsion vehicles (Figure W.5). The term "wye" (or "star") comes from the configuration's resemblance to the letter "Y," where each of the load's components (often windings in an electric motor) is connected at a central point, forming a shape like a "Y." This central point is a standard electrical node, often connected to the neutral point in the power supply [13].

FIGURE W.5 An example of wye configuration.

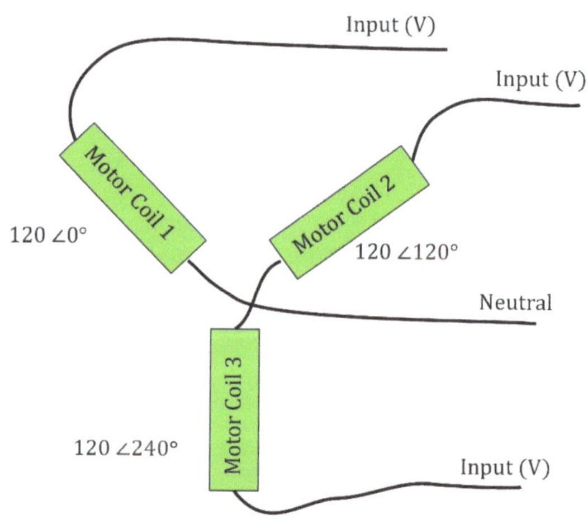

Wye-connected motors are prevalent in EVs and HVs due to their efficiency and smooth operation. In a wye configuration, the start of each winding is connected to form the central point, while the ends are connected to the power source. This setup allows the motor to operate efficiently at lower voltages and provides a phase-neutral voltage for each phase, which can be useful for starting and controlling the motor.

Wye-connected motors are typically used for propulsion systems in EVs and HVs. The wye configuration facilitates a high level of torque and efficiency, which is crucial for these vehicles' dynamic performance and energy efficiency. It also enables the use of regenerative braking systems, which convert the vehicle's kinetic energy back into electrical energy, improving its overall efficiency.

Wye loads in electric motors offer advantages such as reduced phase currents, lower thermal stress on the motor's windings, and improved torque characteristics. This configuration also allows for a more straightforward design of variable-frequency drives (VFDs), which control the speed of the motor by varying the frequency and voltage of the electrical power supplied to it.

REFERENCES

[1]. Society of Automotive Engineers, "J3200 Fluid for Automotive Electrified Drivetrains," SAE Publishing, Warrendale, 2022.

[2]. Society of Automotive Engineers, "J2910 Recommended Practice for the Design and Test of Hybrid Electric and Electric Trucks and Buses for Electrical Safety," SAE Publishing, Warrendale, 2014.

[3]. Society of Automotive Engineers, "J1715 Hybrid Electric Vehicle (HEV) and Electric Vehicle (EV) Terminology," SAE Publishing, Warrendale, 2022.

[4]. Zhou, Y., Li, M., Long, Q., Yuan, D. et al., "Development of a Control System for Permanent Magnet Synchronous Motor Based on LabVIEW and FPGA," *SAE Int. J. Adv. & Curr. Prac. in Mobility* 5, no. 2, pp. 463-472, (2023), doi:https://doi.org/10.4271/2022-01-0732.

[5]. Tang, Y., Paulides, J.J., Kazmin, E., and Lomonova, E.A., "Investigation of Winding Topologies for Permanent Magnet In-Wheel Motors," *The International Journal for Computation and Mathematics in Electrical and Electronic Engineering* 31, no. 1, 88-107, (2012).

[6]. Society of Automotive Engineers, "J1673 High Voltage Automotive Wiring Assembly Design," SAE Publishing, Warrendale, 2012.

[7]. Society of Automotive Engineers and USCAR, "SAE/USCAR-23 Revision 1 - Road Vehicles – 60 V and 600 V Single Core (ISO/Metric) Cables – Dimensions, Test Methods and Requirements," SAE Publishing, Warrendale, 2017.

[8]. Society of Automotive Engineers, "SAE J2501 Round, Screened and Unscreened, 60 V and 600 V Multi-Core Sheathed Cables," SAE Publishing, Warrendale, 2012.

[9]. Society of Automotive Engineers, "J2344 Guidelines for Electric Vehicle Safety," SAE Publishing, Warrendale, 2020.

[10]. Society of Automotive Engineers, "J1654 Unshielded High Voltage Primary Cable," SAE Publishing, Warrendale, 2021.

[11]. Society of Automotive Engineers, "J2183 60 V and 600 V Single-Core Cables," SAE Publishing, Warrendale, 2012.

[12]. Kumar, D., Jain, V., and Potluri, R., "Energy Efficiency of Battery Electric Vehicles with In-Wheel Motors," *SAE J. STEEP* 4, no. 1, 17-37, (2023), doi:https://doi.org/10.4271/13-04-01-0002.

[13]. Society of Automotive Engineers, "J3068 Identification of Vehicles and Supply Equipment for Conductive AC Charging," SAE Publishing, Warrendale, 2023.

X

xEV

Any electrified propulsion vehicle with a high voltage system, including, but not limited to, HEV, PHEV, PEV, BEV, FCEV, and EV [1].

For more information, see **J1715** [1].

REFERENCE

[1]. Society of Automotive Engineers, "J1715 Hybrid Electric Vehicle (HEV) and Electric Vehicle (EV) Terminology," SAE Publishing, Warrendale, 2022.

Z

Zero Emission

See **ZEV (Zero Emission Vehicle).**

ZEV (Zero Emission Vehicle)

A ZEV produces no tailpipe emissions of GHGs or other harmful pollutants during operation. ZEVs utilize alternative propulsion systems, such as electric motors powered by batteries or hydrogen fuel cells, to propel the vehicle without relying on ICEs that burn fossil fuels. By eliminating tailpipe emissions, ZEVs help reduce air pollution, mitigate climate change, and improve overall environmental quality. Common examples of ZEVs include BEVs, hydrogen FCVs, and PHEVs when operating solely on electric power. ZEVs are crucial in transitioning toward a more sustainable and low-carbon transportation system, offering a cleaner and greener alternative to traditional gasoline and diesel-powered vehicles (**Figure Z.1**).

FIGURE Z.1 An overview of the key components of an all-electric car [1].

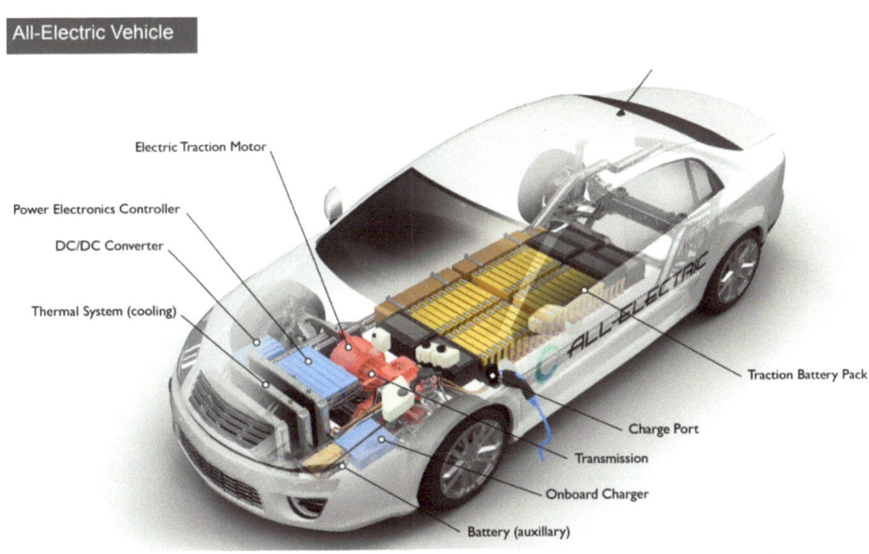

ZigBee Alliance

The ZigBee Alliance, now known as the Connectivity Standards Alliance (CSA), comprises hundreds of companies that create, maintain, and deliver open, global IoT standards. While not specifically automotive-focused, the standards and technologies developed by the ZigBee Alliance, such as ZigBee itself, are increasingly relevant in the context of EVs and HVs, especially for V2G communication, smart charging, and integrated home EMSs.

ZigBee technology can facilitate wireless communication between vehicles and charging infrastructure, smart home devices, and EMSs in EVs and HVs. This allows for seamless data exchange and control, enabling features like scheduled charging during off-peak hours, real-time energy usage monitoring, and optimized energy consumption based on the vehicles and home's energy needs.

ZigBee operates on the IEEE 802.15.4 standard for low-rate wireless personal area networks (LR-WPANs). Its low power consumption and ability to create large mesh networks make it suitable for various IoT applications, including automotive ones. ZigBee can enhance the connectivity and interoperability of different systems within the vehicle ecosystem for EVs and HVs, supporting sustainable energy practices and smart mobility solutions.

REFERENCE

[1]. US Department of Energy, "How Do All-Electric Cars Work?," Energy Efficient & Renewable Energy, accessed June 11, 2024, https://afdc.energy.gov/vehicles/how-do-all-electric-cars-work.

Index

D

E

P

Q

R

S

About the Authors

Courtesy of Jon M. Quigley.

Jon M. Quigley PMP (204278) CTFL is a principal and founding member of Value Transformation, a product development (from idea to product retirement) and cost improvement organization established in 2009. Jon has an engineering degree from the University of North Carolina at Charlotte, two master's degrees from the City University of Seattle, and two globally recognized certifications. Jon has over 30 years of product development and manufacturing experience, ranging from embedded hardware and software to verification, process, and project management.

Jon won the Volvo-3P Technical Award in 2005, going on to win the 2006 Volvo Technology Award. Jon has secured seven US patents and several international patents. These patents range from multiplexing systems and human–machine interfaces to telemetry systems and driver's aids. Jon has been on the Western Carolina University Master of Project Management Advisory Board and the Forsyth Technical Community College Advisory Board. He has also been a guest lecturer at Wake Forest University's Charlotte, NC campus, and Eindhoven Technical University (Holland).

Jon has authored or contributed to more than 20 product development and project management books. His books are used in bachelor- and master-level classes at universities worldwide, including Eindhoven Technical University, Manchester Metropolitan University, San Beda College Manila in the Philippines, and Tecnológico de Monterrey.

In addition to more than 70 magazines, e-zines, and other outlets, he writes three recurring columns:

1. PMTips Quigley and Lauck's Expert Column
2. Assembly Magazine, Ps and Qs on Project Management and Quality
3. Automotive Industries, Quigley's Corner on Automotive Product Development
4. MPUG-Transformation Corner
5. IEEE Reliability Magazine
6. Thomas Cagley's Software Process and Measurement Cast (SPaMCAST) podcast

Jon has given numerous presentations at technical conferences on various domains of product development, including product testing, learning, agile, and project management. He has also frequently been interviewed by numerous business and project magazines, podcasts, and webinars.

Jon is the co-author or contributed to the books:

- Co-author of Taylor & Francis book—*Project Management of Complex and Embedded Systems: Ensuring Product Integrity and Program Quality*, ISBN: 1420070256.
- Co-author of Taylor & Francis book—*Scrum Project Management*, ISBN: 1439825157.
- Co-author of Taylor & Francis book—*Testing of Complex and Embedded Systems*, ISBN: 1439821402.
- Co-author of Software Test Professionals/Redwood Collaborative Media Professional Development Series—Saving Software with Six Sigma, ISBN: 978-0-9831220-0-5.
- Co-author of Software Test Professionals/Redwood Collaborative Media Professional Development Series—Aggressive Testing for Real Pros, ISBN: 978-0-9831220-1-2.
- Co-author of Taylor & Francis book—*Total Quality Management for Project Managers*, ISBN: 978-1-4398-85055-5.
- Co-author of Taylor & Francis book—*Reducing Process Costs with Lean, Six Sigma, and Value Engineering Techniques*, ISBN: 978-1-4398-8725-7.
- Co-author of Society of Automotive Engineering book—*Project Management for Automotive Engineers: A Field Guide*, eISBN PDF: 978-0-7680-8315-6; eISBN prc: 978-0-7680-8316-3; eISBN epub: 978-0-7680-8317-0.

- Co-author of Taylor & Francis book—*Configuration Management: Theory, Practice, and Application*, ISBN: 978-1482229356 (second edition underway).
- Contributor to the e-book—*Opening the Door: 10 Predictions on the Future of Project Management in the Professional Services Industry* with MavenLink and ProjectManagers.net.
- Co-author, Taylor & Francis book—*Configuration Management, Second Edition: Theory and Application for Engineers and Managers*, August 2019.
- Co-author, Taylor & Francis's book—*Continuous and Embedded Learning for Organizations*, August 2020.
- Co-author, Taylor & Francis book under contract—*Risk Management Handbook: A Handbook for Managing Tomorrow's Threats* to Project Success Today Q1 2024.
- Co-contributor of *Scrum Project Management* topic for the *Encyclopedia of Software Engineering*, ISBN: 1-4200-5977-7 and e-ISBN: 1-4200-5978-5.
- Contributor to the e-book—*The Project Manager Who Smiled* by Peter Taylor. Catalog record for the book is available from the British Library ISBN 978-0-9576689-0-4, production June 2013.
- Co-author of Society of Automotive Engineering book—*Modernizing Product Development Processes: Guide for Engineers*, production June 31, 2023, ISBN: 978-1-4686-0541-9.
- Co-author of the Society of Automotive Engineering book—*Dictionary of Test and Verification*, October 2023.
- Co-author of the Society of Automotive Engineering book—*Dictionary of Commercial Vehicles*, July 2024.
- Co-author of the Society of Automotive Engineering book—*Dictionary of Automated and Connected Vehicles*, production under contract.

He has co-authored or interviewed hundreds of magazine articles on various product development, teams, and manufacturing topics in over 60 magazines, e-zines, and other outlets.

Jon enjoys the beauty of nature, hiking in the woods, and playing the bass. Jon lives in Lexington, North Carolina. Email Jon at jon.quigley@valuetransform.com.

Courtesy of Fred Starkey.

Fred Starkey is a US citizen, who completed a B.S. degree in Electronics and Computer Technology at Indiana State University in 1982, and holds three patents.

Mr. Starkey, an electronics hobbyist from a very early age, finished high school with an Indiana Radio & TV service license, courtesy of the local vocational school. He worked as a technician in audio repair shops, commercial audio and community antenna television (CATV) installations, video game repair, and campus television services, while earning his B.S. in Electronics and Computer Technology.

He began his engineering career developing test program sets for linear automated test systems at Delco-Remy. Process failure analysis of the incoming product stream provided the opportunity for advancement into four new areas of expertise: engine controls, thick film modules, robotic assembly, and most importantly, environmental test. Later moving to Delco Electronics Division, Mr. Starkey continued test program set development with an emphasis on socket hardware development for radio mixer and receiver chip test. These assignments renewed and refreshed Mr. Starkey's radiofrequency (RF) background, a skill that would prove crucial. Mr. Starkey shifted briefly to test program set development under naval contracts before returning to Delco Electronics as a manufacturing engineer in the Instruments & Controls group.

Mr. Starkey was recruited by Schrader Automotive early in its development of Tire Pressure Monitoring Systems, tasked with North American original equipment manufacturer (OEM) representation for the tire pressure monitoring systems (TPMS) development team. He was responsible for the development and execution of TPMS qualification and directed the launch quality activities of the first General Motors (GM) and Chrysler programs. Within the development program, Mr. Starkey was deeply involved in design reviews, analysis, and forward product planning. He holds three patents in the TPMS field.

Upon leaving Schrader, Mr. Starkey began a technical consulting practice, which remains active to date. In addition to further service to Schrader, he has enjoyed two particularly long-term clients.

In the first, Mr. Starkey's Electrical Systems Engineering work for Myers Motors helped return the Corbin Sparrow EV to the market as the Myers NmG, prior to Tesla's launch. His work in advanced battery and battery management technology set the stage for Meyer's progression from 3-wheel, single-seat EVs, to 4-wheel tandem models. He built out much of the Myers Point 5 prototype, which may be seen at MyersEV.com.

While Myers pursued 4-wheel, small car solutions, Canadian start-up ElectraMeccanica built the Solo, a clean sheet design based on the Sparrow/NmG concept. When their full production model encountered recall issues, Mr. Starkey solved those issues on behalf of Myers. He continues to support owners of the Sparrow/NmG/Solo vehicles with repairs and battery upgrades.

Starkey was featured at the International Quality & Productivity Centre (IQPC)'s Intelligent Tire Technology Conference for the 2007 and 2008 seasons. His workshops dealt extensively with the validation of direct tire pressure monitoring systems, as practiced by Schrader, which form the basis of SAE J2657.

His second long-term client, tire pressure monitoring start-up RightPSI, recruited Mr. Starkey's services in 2007, on the strength of the IQPC conference presentations. He has directed evaluation and continued development of their award-winning automotive tire pressure indicator since that time. He served directly, as Vice President of Product Development at RightPSI, in the role of Primary Investigator for the development of an aviation version of the company's product, under a Small Business Innovation Research contract to the US Air Force.

In addition to consulting, Mr. Starkey has worked in precision optical metrology for battery manufacturing, automotive parts distribution, and a brief venture into the old style of vocational education that launched his career. He is a regular participant in colleague Jon Quigley's consulting firm, Value Transformations, largely in technical writing, copy editing, and curriculum preparation roles.

His career choices are unduly influenced by opportunities for windsurfing, paddle sports, and open-water swimming. Starkey maintains a well-equipped shop, supporting both professional and personal activities. Capabilities include electronics fabrication and test, general fabrication, machining, and carpentry. Much of this capability is directed into automotive restoration, EV conversions, and maintenance of the "family fleet."

www.ingramcontent.com/pod-product-compliance
Lightning Source LLC
LaVergne TN
LVHW051154210926
848486LV00005B/274